ECOLOGY FOR EVERYONE

Communicating ecology to scientists, the public and the politicians

Edited by

Ray Wills and Richard Hobbs

Published by

Surrey Beatty & Sons Pty Limited

The National Library of Australia
Cataloguing-in-Publication entry:

Ecology for Everyone: communicating ecology to scientists, the public and the politicians.

Includes index.

ISBN 0 949324 84 1.

1. Ecology. 2. Mass media and the environment. 3. Communication. 4. Communication in conservation of natural resources. 5. Communication in science. 6. Communication in biology. I. Hobbs, R. J. (Richard J.) II. Wills, Ray T.

577.

Published December 1998

SUPPORTED BY

Department of the Environment, Sport and Territories
Division of Wildlife and Ecology, CSIRO
Kings Park & Botanic Garden
ESA

PRINTED IN AUSTRALIA BY
SURREY BEATTY & SONS PTY LIMITED
32 Rickard Road, Chipping Norton, NSW 2170

Contents

Contents

CHAPTER ONE

Introduction — why ecology for everyone?

RAY WILLS,[1] RICHARD HOBBS[2] and MARILYN FOX[3]

SCIENCE can be defined as the systematic and objective process of performing observations and testing the soundness of conclusions[4]. As scientists we wish to promote ecological research, and to communicate research results. As ecologists, we aim to apply sound ecological principles to the management of our environment, and promote their application by communicating results by the most effective means. But perception can influence the public credibility and acceptance of ecological research without regard to the quality of the science, and this can impact on the ecological basis of decision making. It is important therefore that when we publish ecological research, we make it accessible to everyone.

What are the most effective ways of doing this? Which parts of society do most want to reach? And do scientists need to go further than simple reporting of results, and become actively involved in advocacy and encouraging societal change? Consideration of these questions is crucial if scientists wish to be relevant and make a difference ecologically for the benefit of everyone in our global village.

At the 1996 annual meeting of the Ecological Society of Australia meeting in Townsville (Ecological Society of Australia Inc. 1996), we convened a plenary symposium which centred on the topic of communication in ecology. This book arose from the symposium, and contains chapters by all the presenters of spoken and poster papers, plus a number of other contributions that were either solicited or offered afterwards. The resulting collection of contributions covers a wide spectrum of topics, all broadly relating to how ecologists relate to each other and the rest of the world.

The book contains a mix of perspectives from people from different backgrounds and with different areas of expertise. These perspectives are often non-overlapping or contradictory, and this reflects the current situation with respect to communication by scientists with other groups and with society as a whole. In addition, the book contains a number of contributions outlining methods of improving the communication skills of ecologists in various ways. The aim of the book was not to provide a cookbook communication primer, as there are plenty of those around. However, the material presented or referenced here will, we hope, provide some guidance for those wishing to develop their communication skills. The final element of the book is a collection of case studies which highlight some of the difficulties and benefits associated with communicating with managers, policy makers or the public.

This book builds on previous contributions in this area in Australia, notably in the volumes edited by Saunders *et al.* (1995), Bennett *et al.* (1995) and Lunney (1992). Our aim with this book is to continue the debate on the role of communication in ecology and provide a range of perspectives on this. It is our perception that effective communication is increasingly vital, and it is our hope that this book generates increasing interest from ecologists in the communication arena.

ACKNOWLEDGEMENTS

We thank the Ecological Society of Australia and the ESA96 Conference; CSIRO Division of Wildlife and Ecology; Kings Park & Botanic Garden; Ansett Airlines and Traveland for financial support of the Townsville symposium. We also thank ESA96 organizers and especially Ross Alford, Lin Schwarzkopf, Robyn Jackson and Kay Bradfield for their logistic support and a great venue.

We thank the Biodiversity Program of the Department of the Environment Sport and Territories (now Environment Australia); CSIRO Division of Wildlife and Ecology; and Kings Park & Botanic Garden for financial and logistic support in preparing this publication.

[1]Kings Park and Botanic Garden, West Perth, Western Australia, Australia 6005.
[2]CSIRO, Division of Wildlife & Ecology, Private Bag, P.O. Wembley, Western Australia, Australia 6014.
[3]School of Geography, University of New South Wales, Kensington, New South Wales, Australia 2033.
[4]Angalo de Haberdasheri, a character in *A Scientific Encyclopaedia for the Enquiring Young Nome*, chose to define SCIENCE as: "A way of finding things out and then making them work. Science explains what is happening around us the whole time. So does religion but science is better because it comes up with more understandable excuses when it is wrong. There is a lot more science than you think." (Pratchett 1990).

Pp. 1–2 *in* ECOLOGY FOR EVERYONE: COMMUNICATING ECOLOGY TO SCIENTISTS, THE PUBLIC AND THE POLITICIANS ed by R. T. Wills and R. J. Hobbs. Surrey Beatty & Sons, 1998.

We also thank the following people for refereeing material: Prof. Ralf Buckley, Dr Mike Calver, Pat Comben, Assoc. Prof. Marilyn Fox, Dr Kris French, Dr Ray Froend, Dr Richard Hobbs, Dr Stephen Hopper, Dr Pierre Horwitz, Dr Ric How, Prof. Jamie Kirkpatrick, Dr Jill Landsberg, Prof. Bert Main, Monica Martinovitch, Barbie McKaige, Dr Tony Norton, Prof. Hugh Possingham, Dr Bob Pressey, Prof. Harry Recher, Dr Denis Saunders, Prof. Rob Whelan and Dr Ray Wills.

REFERENCES

Bennett, A., Backhouse, G. and Clark, T. (eds), 1995. People and Nature Conservation. Perspectives on Private Land Use and Endangered Species Recovery. Royal Zoological Society of New South Wales, Mosman.

Ecological Society of Australia Inc., 1996. 1996 Open Forum and Conference. James Cook University of North Queensland, Townsville 9th–12th July 1996. Ecological Society of Australia Inc., Canberra, ACT.

Lunney, D. (ed), 1992. Communication Skills and the Debate on Conserving Biodiversity. Royal Zoological Society of New South Wales, Mosman.

Pratchett, T., 1990. Wings. Transworld Publishers, London.

Saunders, D. A., Craig, J. and Mattiske, L. (eds), 1995. Nature Conservation 4: The Role of Networks. Surrey Beatty & Sons, Chipping Norton.

CHAPTER TWO

Do politicians need ecologists?

PAT COMBEN[1]

Politicians do need ecologists because:

1. *Homo sapiens "politicianii"* is a vulnerable and often threatened species.
2. Ecologists are uniquely placed to be able to actively participate and influence government policy and the public environmental debate by:
   (i) facilitating educational opportunities;
   (ii) providing factual information;
   (iii) comparing and assessing competing environmental situations; and
   (iv) possessing considerable public credibility.

DO POLITICIANS NEED ECOLOGISTS?

IN Queensland more than anywhere else politicians do need ecologists. May I remind you that it was a Queensland Cabinet Minister who said during the 1970s debate on oil drilling on the Great Barrier Reef "Oil will do the reef no harm because even the scientists will tell you oil floats on water — so how can the underwater reef be harmed?"

More recently there have been developed over the past eighteen months, very exciting plans for a new Queensland Herbarium to house the extensive Queensland collection. Apart from the intrinsic worth of the collection, the Banks and Cunningham specimens and holotypes make it of national importance in terms of heritage as well as science. But this much needed new building appears to have been put on hold with the political comment "Why do old plants need a new building?"

Sixty years ago the great environmental philosopher Aldo Leopold wrote, "Ecology is an infant just learning to talk . . . Its working days lie in the future," (Leopold 1949). Today, ecology is an adolescent, growing and maturing rapidly able to give advice and assistance to our public policy decision makers.

David Suzuki stated the special need politicians, public administrators and society members have for ecology when he wrote, "As long as scientists look at nature in isolated bits and pieces, they can never describe the whole. Physicists have understood that, but by and large, biologists have not," (Suzuki 1990).

Ecologists have a unique opportunity and ability to comment on the "big picture" but in having that opportunity also have a responsibility for the whole picture, not just a small bit of it. Throughout our society, individuals and groups look at problems from their narrow perspective's, whether they are pressure groups or specialist scientists, only politics requires the decision maker to take into account the entire picture, ecology is remarkably similar requiring a scientist to take into account the entire picture.

Harry Recher acknowledged those linkages when, in his Theme Address to the Conservation Biology in Australia and Oceania Conference in 1991, he stated "The goal of conservation biology is the preservation of biological diversity and is as much a political process as a biological one. As conservation biologists, we have the knowledge to achieve our goal, but lack the skills or desire to apply these at a political level. For nations such as Australia, most progress in conservation is driven by non-government organizations, not by conservation biologists. The result is conservation programmes dominated by political considerations, rather than ones based on ecological principles." (see also Recher, this volume.)

Possibly the best recent example of political considerations dominating ecological principles is the Ministry of Senator Graham Richardson. During his time as Minister for the Environment much was achieved for the environment, but it was driven by pressure

---

[1]Wildlife Preservation Society of Queensland, 2/133 George Street, Brisbane, Queensland, Australia 4000.

Pp. 3–5 *in* ECOLOGY FOR EVERYONE: COMMUNICATING ECOLOGY TO SCIENTISTS, THE PUBLIC AND THE POLITICIANS ed by R. T. Wills and R. J. Hobbs. Surrey Beatty & Sons, 1998.

groups and public opinion polls. It was a time of change when the Greenies were brought in from the cold and given a place at the table. But was the best environmental agenda, or the highest environmental priorities addressed? The rainforests were in large measure protected, but where was the leadership from the Minister or the green movement for the grassland parks, or the reptiles of the Brigalow belt or the invertebrates or the lichens or the internal parasites of the red bellied black snake?

Simply put, many green groups influence the agenda by influencing politicians via media campaigns about cute and cuddly creatures attractive to the media and public. A little unscientific but emotive language also assists on the path to that influence. The end result can be the wrong set of questions asked, a blurring of the distinction between welfare and conservation questions, a malaportionment of funds towards cuddly creatures with big brown eyes away from less easily loved and identified creatures in just as great, or greater, need. In a nut shell, try running a public support campaign for the internal parasite of the red bellied black snake, when the Koala Foundation has mobilized a campaign "To save the Koala from extinction."

To overcome this bias in media and some sections of the environment movement, there is a need for a strong presence and voice from the professionals in the field, such as yourselves. Whilst this points to ecologists needing to be more politically aware it also points to a symbiotic relationship with a mutual need for truth and honesty. Ecologists need politicians because politicians represent the political system controlling taxation, research monies, the land tenure system providing our reserve system and the inadequate regulatory system which attempts to protect the environmental processes. However inadequate these things may be, they are the tools needed to ensure research and conservation of species diversity.

To ensure the debate and then decisions are correct politicians need ecologists. Politicians firstly need your technical advice to ensure the right decisions are made for the environment, secondly they need your support for the acceptance and promotion of the right answers to the public and pressure groups and thirdly politicians and public both need you to keep the other group of bastards honest.

Ecologists must be major providers of advice for five principal reasons: ecologists are researchers and can have answers, ecologists attempt to look at the total picture, ecologists have public credibility and can be educators, ecologists can be powerful advocates for that which they study, and ecologists can be unbiased.

It is in each of these roles that your friendly politician needs you. Even the politician that has not got a clue about the issues is not a lost cause, but rather an education opportunity awaiting a teacher. It is just like layers of rock, some are a bit harder, tougher and thicker than others, but they can all be used.

Ecologists should not be wary of entering the public debate, particularly to present facts. I once sat in my office being abused by a person well known throughout Australia for her work with koalas. I tried to say I wanted to spend The Department of Environment's limited research money on the needs of a number of the small macropods and not koalas. I was not allowed to say anything of the kind and the lady left me to attend a press conference in which she said I had no understanding of the environment and I would allow koalas to become extinct in ten years. A month later she was in the USA stating there were no koalas in any Australian national park. Where were the ecologists there?

How do ecologists become part of the decision making system? To influence the political and public debate on environmental issues there are at least three major hurdles to overcome. Firstly (drawing on the characterizations in the BBC television series "Yes Minister"), to convince Jim Hacker (the Minister) and minders that a problem exists at all, then to convince Sir Humphry Appleby (the Director General) that a careful review to examine how a public inquiry into what the terms of reference for an investigation into the jurisdictional question for a preliminary inquiry is perhaps not the way to go, and finally to convince the public (in all its facets) of the benefits of change. This later suggestion of influencing public thought can be at the time interest groups are forming policy or later in advocacy for a given policy. No group is better placed to undertake those roles than ecologists. In political terms, in the advocacy of the ecologist lies the hope of the world.

My experience as a minister led me to the conclusion that there was an abysmal ignorance of environmental matters within my colleagues. They were not quite philistines but there was generally a focus on the next edition of the Courier Mail or the next election. If there is no appreciation of niches and habitats then they will say "Why can't we move that ground parrot from the swampy area and put it in a tree with the other parrots Pat?" True story. In this there is an opportunity to offer advice and assistance to improve the debate.

Most other major areas of endeavour such as sport, manufacturing, education etc. have professional bodies well tuned to lobbying. It seemed to me that often the environmental debate was left to amateurs. Often this has excellent results, but not always. Too often public perception and not ecological worth led the debate.

It is not easy to have that influence. It would be difficult to get back to the days of Sir Joseph Banks presiding with the full pomp and ceremony of a Magisterial President of the British Royal Society that informed government in a very formal manner of the governmental policies to be followed concerning the natural sciences and exploration. Or in more recent times in Queensland the sort of influence exercised by the National Parks Association of Queensland in the 1950s and 60s when no national park was created without the Association's imprimatur and it was the acknowledged conscience of government. It is difficult to produce timely material when other less qualified groups have staff paid from the public purse, media contacts as well as a certain level of public credibility. But that is the current system and the one within which you must act.

Your immediate advantages are in the area of credibility. Ecologists, particularly the Ecological Society of Australia have a huge corporate knowledge which can be put to public use. When speaking as a Society it has an independence of government and motive and bias that few individuals or conservation groups could have.

May I therefore make my first suggestion to you that a way be found to allow your knowledge and credibility to be available to governments. Whether it is background papers, or an offer of independent surveys of a panel of independent experts, some formal mechanism should be found. Why is there no formal government liaison sub-committee of ESA? Or a Conservation or a Funding Advocacy sub-committee. To get results you need to influence the decision makers and to do that on other than crisis management or *ad hoc* basis, you need a structure.

Secondly get involved at a formal level with some of the major conservation organizations. This will not always be easy because they will be in perpetual crisis mode and not be receptive to steadying hands. From my personal experience the leadership of your member Dr Roger Kitching of the Queensland Conservation Council has been of great benefit as have the writings of Denis Saunders of Canberra, which allow an amateur to understand the arguments and carry them to the political debate.

Thirdly do your homework — find out who the decision maker really is. I will leave it to Peter Bridgewater (see Bridgewater and Press, this volume) to distinguish the bureaucratic image but the best example I personally recall was during the early days of the Fraser Island Campaign. About half the island was national park, the other half was forestry and still being logged. As a campaign commenced to stop the logging I (as then Environment Minister of Queensland) was inundated with letters demanding I stop it. When I raised the matter in Cabinet, the Premier said "That's not your area of Ministerial responsibility Pat, Ed Casey what have you heard?" Ed Casey, the Minister for Primary Industries, responsible for logging, said "Not one letter Premier." Who then said to me "Does not sound much of a problem Pat". The campaign to protect Fraser Island, which I totally supported, was set back six months by targeting the wrong person.

Fourthly do not be scared to approach individual politicians and offer assistance from time to time. It will surprise many of you the extent of ignorance in our elected representatives. But if properly approached they will generally say "yes" to offers of assistance with a speech on a given piece of environmental legislation.

Let me finally suggest three ways in which this Society could be unique.

1. Occasionally pen a letter to a politician who has actually made a courageous decision and say "well done". I have still got both the "thank you" letters which I received in my six years as a Minister.
2. Put a bit of backbone into arguments for things with no backbone. A News Release or report about the little critters as compared with the cute and cuddleys should get media attention and eventually focus on the full scale of ecological needs.
3. Use your professionalism to keep the discussion and debate honest, it will be a novel approach in an environmental debate in which the power of money is too often against the passion of the environment.

CHAPTER THREE

The voice of reason: science and technology talking to government

TOSS GASCOIGNE[1] and JENNI METCALFE[2]

How certain does a scientist have to be before offering advice to government? On the release of RCD, or the effect of logging old growth forests, or whether boats should be allowed to anchor in the Great Barrier Reef? We conclude that scientists expert in a particular field must be prepared to enter contemporary political debates and offer opinion to the decision-makers, if they are ever to throw off that "ivory-tower" label. We will examine some of the barriers and incentives which inhibit or encourage scientists to become involved in issues, and look at appropriate mechanisms for affecting the outcomes of environmental debates.

INTRODUCTION

LET us accept that scientists have a role and an interest in affecting the course of governments, and that they are prepared to voice opinions in controversial areas where economic interests may clash with each other or with environmental interests. Let us also accept that they are prepared to offer advice in an area of their expertise, even though a particular subject may be clouded by uncertainty or incomplete experimentation.

These are fairly large assumptions, particularly the third one. Scientists and technologists inhabit a world different to most others. It has a concern for truth, for proof, and for accuracy which is not shared by other professions (such as economists, for instance). The scientific world has a modesty about expressing opinion, especially before the ideas have been tested in the realm of their peers through publication in peer-reviewed literature.

But governments have to govern, and the managers have to manage. If scientists are unable to provide opinions and advice about the consequences of certain actions, then government will seek it elsewhere. Often these alternative sources will be far less qualified to comment than the scientists and technologists who felt unable to comment from what they see as a proper commitment to accuracy and proper scientific procedures. The government, on the other hand, views this as unhelpful, unworldly and rather precious.

But the assumptions have been made. You are ready to wield influence. How, then, should you go about helping set government policy?

The first rule is to think of the people you are trying to convince. Who are they? What do they want? What do they know already? Will they be familiar with technical terms? Do they have the expertise, the time or the interest to read detailed material? What options do they have in mind?

Generally, simple material is most effective with these audiences. Get to the bottom line quickly — and remember that policy makers (like most people) are more interested in the effect of your results than your clever science.

Before you launch into your own exercise in moving the direction of government, remember that FASTS — the Federation of Australian Scientific and Technological Industries — is the peak voice for working scientists and technologists in Canberra. FASTS talks to government and the bureaucracy continuously, and can assist Member Societies to make the right sort of impact in Canberra.

Here now is the do-it-yourself guide, five tactics to move government. Consider using them in combination on any given issue — and remember that persistence pays.

THE MEDIA

Everyone reads or watches the media. Politicians make decisions on what they see on television news, and ask questions in the House on what they read in the paper. Use the media as one arm of your strategy, to open doors, help forge alliances, get your issue talked about.

Making the media work to your advantage requires imagination, lateral thinking and deft

[1]FASTS, P.O. Box 218, Deakin West, Australian Capital Territory, Australia 2600.
[2]Econnect, P.O. Box 784, South Brisbane, Queensland, Australia 4101.

Pp. 6–8 *in* ECOLOGY FOR EVERYONE: COMMUNICATING ECOLOGY TO SCIENTISTS, THE PUBLIC AND THE POLITICIANS ed by R. T. Wills and R. J. Hobbs. Surrey Beatty & Sons, 1998.

footwork. Use experts to advise you. How do you write and distribute a media release? How do you contact journalists? How do you stop them from getting things wrong?

The great difficulty for scientists and technologists is that you have to learn a different language to succeed in the media. The electronic media in particular does not have time or the audience to give reasoned, detailed treatments of issues that carefully weigh the merits of both sides.

If you are on television news, expect your story to run for one minute 20 seconds, and for you as the main spokesperson to comment directly for about nine seconds. Now you are in the world of "infotainment"; and if your story is not brief, interesting and accompanied by dramatic pictures, it will not appear at all — no matter how worthy or important it is.

ABC journalist Pru Goward said in her Senate Occasional Lecture Series talk in 1995: "the information which works best on radio and television has to be simple . . . Story-telling is the best vehicle for transferring information . . . Any good salesman or politician on television avoids conceptual words, report-writing words and unfamiliar words at all costs. You cannot think and take in at the same time."

PERSONAL REPRESENTATION

Go and talk to government. Make an appointment to see the relevant Minister — or you may have to be satisfied with one of their staff (each Federal Minister has half a dozen staff members).

Precede the meeting with a letter setting out the agenda, and keep the discussions simple. Follow the meeting with a letter summarizing the main points, and the agreements and actions reached.

Try to be helpful. The two biggest mistakes deputations to government make are asking for too much, and raising problems without offering solutions. Do not have more than two (perhaps three) members in your deputation. Rehearse what you are going to say.

Ministers like to help but they are driven by many conflicting pressures. Do some homework so you know a little about them, including where their electorate is and what their interests are. (Former Minister for Science and Technology Peter McGauran lists his as horse-racing, ballet and international law; Minister Moore's are tennis [he played Wimbledon 1956], cricket, golf and reading.)

To be effective, the contact has to be ongoing. You will not achieve all you want in one meeting. Often it is a matter of being viewed as a potentially helpful source of advice and ideas; and in this way, your views will gradually accumulate with government.

David Kindon, a Canberra PR consultant, carried out an illuminating survey of politicians. He asked them how they preferred to be approached by lobbyists. The results are published in a helpful nuts-and-bolts "Guide to Politicians' Lobbying Preferences" available from his company (P.O. Box 4893, Kingston ACT 2604) or free on his WWW site.

FORUM FOR BUREAUCRATS

Bureaucrats are highly influential in the formulation of policy, so make sure they are well-informed. For instance, FASTS is convening a discussion of the new FASTS policy document involving a handful of mid-ranging staff from the Department of Industry Science and Tourism (DIST) and several representatives of FASTS.

The main task lies in the identification of the right government departments and the appropriate people within those departments. Working out who the best people are will require some preliminary work, and you will need to use departmental contacts to lead you to the right people. The most efficient method is probably to persuade some higher ranking officer to convene a meeting on your behalf.

Another approach is to invite officers in charge of a particular policy area to visit an experimental site. Some bureaucrats may feel quite divorced from practical field work, and would appreciate the chance to see an experimental site and discuss the work with the scientists and technologists in charge.

Bureaucrats have to fit into structures and programmes, so in framing your opening remarks, think about what is going to be useful to them. Keep it relatively straightforward at first, and allow plenty of time for their comments and questions. Offer yourself as a resource of ideas, because successful lobbying depends on maintaining day-to-day contacts, and bureaucrats as well as politicians need experts they can consult on various matters.

ALLIES AND NETWORKS

Small groups are generally easy to ignore, so find other groups which agree with you on a particular issue and are prepared to make joint representations. One weakness in science and technology is the number of small or mid-sized groups (with equally small budgets) which represent interests in this area.

Alliances and networking are one of the keys to successful lobbying, so get organizations like the National Farmers' Federation (NFF) and the Australian Conservation Foundation on side. Make up a joint delegation to discuss the issue with the Minister.

The value of alliances is considerable. Wendy Craik, Executive Director of the NFF, sees them as an increasingly important part of the lobbying scene. In an article in the Canberra Times (June 20, 1996), she said: "If we link up with the miners and the foresters on an issue we are representing 75 per cent of exports and that gives us significant clout when talking to the bureaucracy or the government.

"The tariff concession issue proved this. There were 30 industry organizations who presented the government with a united view on the issue. The result was that we achieved a lot of what we sought but if each association had approached the government individually the results would have been a lot worse for us."

These alliances cannot be forged overnight, so start building the links early.

THE GRASSROOTS

The local constituency cannot be ignored, and at least one major lobbyist says that there has been a shift in recent times to spend more time on grass roots work and establishing relations with community groups and regional people.

Robert Bain, Executive Director of the National Association of Forest Industries, said in the same article: "Effective lobbying groups have to have some links back to the community or the community constituency. The idea that lobbying is being done by smartly turned out high-profile Canberra individuals who can organize meetings is over".

"Industry people see the process of contact with government as more of a day-to-day, nitty-gritty affair by someone who has contacts."

CONCLUSION

The task of speaking to politicians and government is an important one. They have shown little evidence over the years of understanding how scientists and technologists operate, or how their abilities can be harnessed to make the greatest contribution to the economic, social and environmental benefit of Australia. And while government's understanding is limited, so is its willingness to confront the problems scientists and technologists face in doing their bit for Australia.

CHAPTER FOUR

Public and political: the challenge for ecologists

HARRY F. RECHER[1]

Collectively scientists accept responsibility to further human welfare through education and research. I contend that scientists are also accountable for the welfare of the planet and the other organisms with which we share Earth. Scientists can only exercise accountability by assuming responsibility for the use of the knowledge and technology they create. This means scientists must accept that education includes communication with people outside the confines of the formal education system, and that there is a requirement to ensure that knowledge and technology are used in environmentally and socially responsible ways with regard to intergenerational and interspecies equity. The scientific community in Australia has largely failed to do this. Consequently science and education have been marginalised and the products of scientific inquiry used with little regard to social and environmental consequences. While changes are needed in the way scientists are trained for science to meet its responsibilities, there is also a need for scientists to be involved in the political process where decisions on the use and development of scientific knowledge are made. An involvement in political processes may be a special duty of scientists who call themselves ecologists and who therefore take pride in being able to see humanity as part of a global ecosystem. Yet, political involvement is rejected by most ecologists. There is a frequently presented argument that ecologists (scientists) lose credibility through political involvement and that politics is not the role of ecologists or of their professional bodies. I disagree.

INTRODUCTION

WE manage an incredibly complex society that is embedded in an equally complex biological system. This is done with an incomplete knowledge and understanding of our actions and their consequences. Our knowledge and understanding are incomplete because we lack some information, but as a society we ignore and suppress much of the information we do have information which is required for managing complex biological systems (ecosystems). As ecology is arguably the most important of scientific disciplines, ecologists need to acknowledge their social responsibilities and make a better effort at increasing our knowledge and understanding of ecological systems, while striving to ensure that the information that is available is used, and used to best advantage. Ecologists will only meet these goals by dropping the traditional boundaries of science.

What is our individual role as scientists and ecologists in this community of people that we live in? My own role is clear to me: science as a whole, and ecologists in particular, must be publicly involved in the decision making or political process. As part of my commitment, I often speak against the failure of any but a few Australian ecologists to be involved or to show anything other than the most marginal of intentions to become involved with our social and political systems. I am particularly angry at those who profess a concern for other organisms, but who fail to promote the conservation and survival of Australia's rich biodiversity. Most ecologists acknowledge that the conservation of biodiversity is important, but few take a public position to protect biodiversity against our societal commitment to and idolization of growth and development. My own view is that growth includes the issues of immigration and the presumed right of people to have children. Ecologists are uniquely trained to advise and direct the choices made by the community on these issues. I reject the notion that as much, or more, can be achieved by working quietly behind the scenes. While advising governments and "whispering in the ears of Ministers" has its role, we only need to look at the failure of governments to respond adequately to environmental issues and at the loss of continental biodiversity to see that whispering achieves little and is commonly ignored. It can be ignored, because what is whispered is not public knowledge.

What is lacking among ecologists in Australia is a professional sense of responsibility — an ethic which would be a guide to our behaviour as scientists and as ecologists. Ecology and ecologists have important roles to play in society over the next few decades. No one expressed this clearer than Aldo Leopold when

[1]Department of Environmental Management, Edith Cowan University, Joondalup, Western Australia, Australia 6027.

Pp. 9–15 *in* ECOLOGY FOR EVERYONE: COMMUNICATING ECOLOGY TO SCIENTISTS, THE PUBLIC AND THE POLITICIANS ed by R. T. Wills and R. J. Hobbs. Surrey Beatty & Sons, 1998.

he singled out ecology as a discipline whose "working days lay in the future" (Leopold 1949). Leopold's writings and his portrayal of a "land ethic" have given him the stature of a leader among people, but after 50 years there is little evidence that ecologists have done much more than admire his wisdom and lament his passing.

As I have argued before, the traditional roles of ecologists have been in education and research; these are comfortable roles (Recher 1990a). They are the roles most ecologists act out and have been trained for at university. Despite the comforts of education and research and their benefits to society, it is no longer acceptable for ecologists to be content with these roles in society. This is not an ideal world. The threats to the global environment and the risks to the survival of many (if not most) of the world's other species do not give us time for inaction.

If you think I am being a Jeremiah and doom-sayer, then consider that I travel in illustrious company; twice in 1993 the world's scientific academies and leading scientists issued public declarations of concern for the survival of humanity and the planet unless action was taken to reverse population growth and solve global environmental problems (Ehrlich 1996; Ehrlich and Ehrlich 1996). The publication of these statements does not contradict my concerns about the failure of scientists to be publicly involved. Issuing public statements, signed *en masse*, is politically important, but it must be followed by continuing public involvement by individuals and professional bodies. Single statements, regardless of the prestige of the signatories, are like the whispers in the halls of power — easily pandered to, easily forgotten. As Ehrlich (1996) laments, neither petition by the world's scientific community received significant media attention. Involving the media is critical in getting messages to governments and the people and requires individual action.

If global biodiversity is to survive the next century, then ecology and ecologists must become advocates of change. They must not only advocate ecologically sustainable development, but they must promote values outside the economic framework of traditional society. Ecologists need to be outspoken critics of society and its short-term aspirations for material wealth. A long-term and balanced approach to our use of resources must be argued. For resources that are already over-exploited, such as old growth eucalypt forests, inland waters and coastal estuaries, a balanced approach does not mean sharing the residues between conservationist and those who would continue their exploitation and development. It means recognizing and arguing that the extent of past use means that no further use is possible until systems recover. Criticism and argument must be based on research, but in the absence of data, ecologists must argue from theory and to be prepared to make predictions. We cannot afford the luxury of statistical perfection. We should not risk our credibility, but also we should not sacrifice our world by adhering too rigorously to scientific methodology. As Paul Ehrlich puts it, 'we cannot stand aside while our laboratory and home is vandalized'. Nor should we leave it to environmental organizations to set the agenda and present the arguments.

ENVIRONMENTALISM

Environmentalists and environmental organizations often have a narrow agenda (e.g., ending woodchipping, saving wilderness, stopping hunting). While environmental causes are often praiseworthy, they are not necessarily the most important issues and tend to drag political attention away from more serious, but less charismatic, problems (e.g., loss of woodlands, degradation of inland waters, overgrazing of pastoral lands). For example, the emphasis placed by environmental organizations in Australia on wilderness conservation has made it almost impossible to achieve a mix of forest conservation and management that might ensure the survival of forest biodiversity while creating an ecologically sustainable forest products industry.

Wilderness parks have biodiversity values, but these are not necessarily representative of forest ecosystems; all too often, wilderness parks are established on unproductive soils or at high elevations (Hall 1988; Resource Assessment Commission 1991). However, the large areas of wilderness reserves (e.g., the forest parks on nutrient poor sandstones in central New South Wales) create an illusion of effective conservation and allow forestry interests to press for more intensive logging on the forests outside the reserve system.

A representative forest reserve system needs to include large areas of forest on the rich and more productive soils at lower elevations. Such forests, by virtue of their productivity and ease of access, have a history of logging and disturbance by people. Restoration may be necessary to restore ecosystem functions and processes (Lunney 1991). Although such forests do not qualify as wilderness, their importance in biodiversity conservation is extraordinarily high. In addition, it is both necessary and possible to accommodate biodiversity conservation within areas being harvested for timber (e.g., see Shaw 1983; Recher *et al.* 1987; Davey 1989; Scott 1991); without such an approach, the long-term maintenance of the ecological integrity of forest ecosystems is unlikely (Norton and Lindenmayer 1991). The prevailing system of allocating forest to either inviolate reserves with an emphasis on wilderness (promoted by environmental organizations) or to intensive harvesting (promoted by forestry interests) may be the worst possible choice for the long-term conservation of forest ecosystems and their associated biota. Instead, the design of a reserve system, while considering emotional and aesthetic arguments as well as commercial interests, must be based on ecological principles (see Recher 1994). Achieving this means that ecologists must have a more prominent role in conservation decisions.

Ecologists have been trained — perhaps they were already so inclined — to view the whole and to see and understand the interactions between all parts of the environment. More than any group of people, ecologists can look to the future and see the consequences of even trivial actions in the present. In this they are unique. This is why ecology is the most important of the sciences. While ecologists need to maintain their roles as educators and researchers, they have responsibility for the education of the community as a whole and need to extend their research to the active promotion of new management guidelines for the conservation and use of the world's resources.

Too many ecologists are too silent. There are reasons for this which I will discuss, but the reasons are inadequate and unacceptable.

ECOLOGIST AGAINST ECOLOGIST

I am concerned about the failure of ecologists to respond publicly and to be involved with decision making at a public level. I am even more concerned when I encounter ecologists and other scientists speaking publicly in favour of exploitation, development, and unending growth. Such views have become more common in recent years as commercial and anthropocentric elements in society react against the constraints on growth and development imposed by regulations designed to protect and enhance environmental quality, and to protect other species (Ehrlich and Ehrlich 1996). As Ehrlich (1996) notes, a central theme of the advocates of unrestricted growth and opponents to environmental management is that greenies are extremists. This "brownlash", as the Ehrlich's call it, is as common in Australia as it is in North America. On both continents, the denial of environmental problems and their dismissal as trivial and best corrected by market place forces is accompanied by misinformation (including the deliberate withholding of information and distortion of the truth) (Ehrlich and Ehrlich 1996). Frequently, it is ecologists and other scientists who present these distortions and who prevent information on the state of the environment from being made public.

This being the case, I ask why ecologists who take a pro-development/pro-exploitation stance on natural resources can speak openly and freely, while those scientists with an alternative set of values are conspicuous by their silence or are conspicuously silenced. This includes our professional bodies. We appear to have a situation in Australia where one group of scientists/technologists speaks openly, while another is intimidated, censored, and repressed. Martin (1981, 1992) provides examples.

I am not the only one asking these questions. Jamie Kirkpatrick (1996) in his President's Report to the ESA commented that there are often conflicts between ecologists in conservation/development debates. As an example, Kirkpatrick referred to the exchange in the *Western Australian* newspaper during 1996 between a senior scientist from CALM (the Western Australian Department of Conservation and Land Management), and Susan Rhind, a doctoral student from Murdoch University, concerning the adequacy of forest wildlife management practices in a logging area. Although the exchange arose from a press report of a paper that Rhind had published after peer review in a reputable scientific journal, the *Western Australian Naturalist*, it was alleged that Rhind had allowed her values [protecting the forest from logging] to ". . . overshadow the need for objectively based data." Such allegations are difficult to substantiate, and one must question the need to introduce them into an exchange of views when all that might have been required is a correction of any errors of fact.

A CODE OF PRACTICE

We should ask how is it that ecologists who question the "official" position are subjected to public and other attack by fellow scientists (as well as by others), while we seldom see the reverse. For example, I have never seen a public challenge to the oft expressed view by some ecologists that creating new habitats by dredging sea grass beds or logging forests is beneficial because it increases diversity. Nor have I seen attacks in the media by ecologists on those scientists who promote the view, often without data, that logging has no long-term adverse effect on forest fauna. Perhaps, it is simply not newsworthy.

Kirkpatrick asked how the ESA should respond to public disputes among ecologists. He made several suggestions how the ESA could respond, while recognizing that among the membership there would be different views on any given issue. He also said that it would be unreasonable to expect the ESA to act as an arbitrator in a debate between two ecologists.

Kirkpatrick did say that there should be a code of conduct for debates and it should have the following elements:

1. opinions given should be honestly held and their limitations made clear [to the public]. The basis for opinions should be stated.
2. there should be a separation of personal opinion on the issue and ones' opinion of the character and values of ones' opponents.
3. proponents should communicate with each other.
4. any debate should be open and between the proponents by means of the scientific literature, verbal debate, and the media. Open debate implies the absence of threat and intimidation.

My interpretation of Kirkpatrick's comments is that he advocates that professional bodies, in this case the ESA, need to be involved in social and political issues, if only to keep the participants honest. Thus, the ESA could play the role of an honest broker, but where necessary should correct errors of fact or omission and ensure the public was aware when differences were based on opinion and were not necessarily factual. I also hope that the ESA and similar bodies would respond with public censure when attacks on and threats to individuals exceeded the scientific merits of the case and when they were politically motivated or sought to intimidate. It is important for ecologists, especially those at a vulnerable stage in their career, to know that their professional bodies stand behind their rights and obligations to inform the public on environmental issues. This would not remove the responsibility of individuals to restrict their comments or criticisms to their areas of expertise and to be certain that the views or opinions expressed had a solid factual and theoretical framework for support.

I am reminded of the policy that prevailed during the years that I worked at the Australian Museum in Sydney. Museum scientists were free, even encouraged, to enter into public debate or issue comments so long as it was in the individual's area of expertise and so long as they were correct. Over a twenty year period of public involvement, I never once failed to be supported by the Museum, regardless of possible political consequences. I believe that not only did the public benefit, but that the Museum was a better and more credible institution for the support it gave its scientific staff on controversial issues. Adopting Kirkpatrick's suggestions would make the ESA equally credible and give it strength within the scientific community and among the public.

THE THINGS WE VALUE

Like Gough Whitlam's definition of the National Estate as "being those things that we value", and should therefore protect, my concerns are for those things that I value, and which I am committed to protect. I am aware of contradictions in my views; what I hold passionately to be important, another may not. People, including ecologists, have different value systems. Whereas I hold freedom of choice and opportunity highly and consider the rights of other organisms to survive as equal to the rights of people, another may not. Should I allow each person to express their views and values and not demand that they adhere to mine? Surely individuals are entitled to remain aloof from public involvement — they may be content with the course of human history and their role in it. Many may hold human life dearer than the survival of another species. From listening to those who testify in court in support of environmental destruction or from reading environmental assessments that conclude that development will have no long-term, adverse environmental impact, it is obvious that many scientists must feel this way. I presume, of course, that their testimony and environmental assessments are sincerely and honestly held. Otherwise why would they say what they do?

Ecologists need to address two issues:

1. why do so few speak for the world?
2. why do so many speak against the world?

Is this simply a dichotomy of values or are there more sinister reasons for this division?

Science is supposed to be rational and value free. According to the paradigm that seems to dominate us, science should be removed from politics and detached from the social implications of its endeavours. However, as Seager (1994) points out, this is not possible — science and scientists are not value free. They cannot be. Their research and how they apply the results of that research are fundamentally affected by their personal values and by the value systems of the society within which they live. I do not expect scientists to be value free. I am not independent of my values. My values not only determine my personal agenda, but they determine the nature of the research I do. I ask questions about the things I value and look for answers which will help to conserve and manage those values. I suspect many, if not most, ecologists operate in this way.

Seagar (1994) says that science can and must play a role in environmental management. However, she continues that the nature of scientific inquiry and the way science is applied in technology means that the role should be limited. Seagar argues that the paradigm of science — pursuit of truth and objectivity — is not appropriate for environmental understanding. Knowledge of the environment must consider the social, political, and emotional contexts within which environmental issues arise and are resolved. To some extent, Seagar's views are coloured by her feminist stance and her notion that men (who, she says, dominate science) cannot share or express the same emotions and values as women. I find that view myopic, but agree that scientists (male and female) need to step outside the bounds of the dominant scientific paradigm and make more use of the social, political, and emotional contexts surrounding environmental issues. Are there reasons why Australian ecologists find this difficult? If there are, can we change them?

Does it matter? It matters a great deal. Australia has major environmental problems for a host of reasons (see State of the Environment Advisory Council 1996), but one of these is the failure of the scientific community to do what they are trained to do — and that is to question. We should not only question scientific results, but we should question how the results of science are used.

SIMPLE SCIENTISTS

The environment is a major concern of Australians. During the first half of the 1990s, in poll after poll of public concerns, only unemployment was considered more important in the short term, while the health of the environment dominated traditionally important issues such as the economy, education and welfare in people's perception of their long-term well-being. It is evident that concern for the way we use and manage our natural resources as well as our awareness of environmental degradation and the need for more effective conservation programmes has been incorporated into the way society functions. The issues will change, but the environment will remain on the political and social agenda through this decade and into the 21st century.

It would be nice to say that public knowledge of environmental issues arose from the efforts of ecologists, but the development and growth of the environment as a political and social force for change in Australia is a product of the media and the advent of mass communication. Australian ecologists have mainly played cameo roles, while the centre stage has been held by environmentalists and environmental organizations, such as Greenpeace, the Wilderness Society, and the Australian Conservation Foundation. If ecologists decide to play a larger and more political role in environmental issues as I advocate, they will need to work with the media and learn how to communicate. The presentation of environmental events in simple English suitable for use in the media has been and will remain fundamental to the development of an environmental ethic and to the understanding of environmental issues by the public. Most people have not been trained in the sciences and only a minority have tertiary education. Simple English, clear expression, and a brief message are the hallmarks of effective communication with the public.

Unfortunately few scientists make an effort to communicate either with the media or with the public. There are those who, like Tim Flannery and Paul Ehrlich, are outstandingly good communicators, but when others try, they are unable to put their message in a form that is easy to understand. Again, there are two issues:

The first is the failure (the unwillingness) of scientists to communicate the results of research directly to the public. The way to do this, of course, is through the media in the broadest sense and not just via elite science programmes geared to an educated audience.

The second is the difficulty most scientists have when trying to communicate in simple language; many cannot communicate at all.

- There are reasons why scientists have difficulty in communicating: Many Australian scientists are prevented from speaking publicly or with the media by restrictions imposed by their employers. Others are afraid that their public comments will affect their careers, jeopardise research funds or lead to restrictions on their research (e.g., obtaining permits to work in particular areas). These concerns are especially apt in the context of the environmental debate where many issues are emotional and have political and economic implications and have been explored at length by Brian Martin (e.g., 1981, 1992).

The fears are real. Many Australian scientists who have taken a public position or provided information on controversial environmental issues (e.g., logging, water pollution, misuse of agricultural chemicals, environmental threats to human health) have experienced the jibes and retributions of employers and colleagues. Martin (1981) describes a cabal of the "cognitive scientific elite" and the "political scientific elite". These groups overlap and influence government and corporate policy on a range of subjects from research priorities to the hiring and dismissing of staff. It is not in the self-interest of these scientific elite to challenge the system. It is also in the interest of the elite [much like the religious hierarchy in keeping the masses ignorant during the Dark Ages] to promote science and the training of scientists as it has been done in the past.

- Articles written for lay people or lectures to public (non-professional) audiences are given little value or weight in decisions about promotions and research support. Ehrlich (1996) expresses the contrary view that "... contributions in [public education] should be professionally highly rewarded." My own institution (Edith Cowan University) strongly supports this position and is setting a standard for other universities to aspire to.
- I have said this before (Recher 1992), but most scientists who are educated at Australian (or British) universities receive little or no instruction in communication. They enter university with secondary school level verbal skills (which are poor), and they receive their Ph.D. with about the same level of expression, albeit smoothed by maturity. On the way to their degrees, they acquire the language of the field of their specialty and to survive (pass) they learn to use the jargon of their field instead of the simple English they and their supervisors are incapable of writing. After all, it is only necessary that they be understood by other specialists in the same field.

  A consequence is that a rejection of communication develops (to the point that few scientists even attend departmental seminars outside their area of specialization) and accompanying this, a disdain for communicating with the public. Since they do not know how to communicate, in self-justification, many scientists deride the efforts of others and belittle the media.
- There is also a related problem associated with science education at universities. Not only are young scientists denied the opportunity to develop their communication skills, but they learn little about the society in which they live and work. They take no arts or humanities. There is no training in foreign languages. Little is taught about ethics, religion or aesthetics. Science in taught and practiced in a vacuum that has no history, no humanity, and no emotion. It is therefore not surprising that so many scientists find it difficult to relate their work to the human community or even to understand that others might be interested in what they are doing. To them giving an interview becomes an ordeal. Obviously, we must change the way science is taught at university.

The converse is also true. Remarkably few politicians or members of the corporate boards of major Australian companies appear to have any training or understanding of the ecological sciences. No wonder, perhaps, that so many of our politicians show no interest in, concern for, or understanding of environmental and conservation issues, much less of the needs and value of science and education. So when we change the way science is taught, we also need to change the ways in which accountants, lawyers and doctors are taught. The world is too frail to do otherwise.

Our universities need to move to a common first year in which science students are taught how to be human, and arts students are taught the culture of science and its importance to human endeavour. This will mean that university degrees will be extended to four years for all students and the now irrelevant Honours year will be related to the history books where it belongs. In just the last 50 years, less than an average Australian's life span, the expansion of knowledge has been huge. We delude ourselves to think we can educate either science or arts students in three years at university. In 1996, a Bachelors degree is little more than equal to a High School degree in the 1930s. Postgraduate studies through the doctorate level should now be viewed as the equivalent of a university degree of only one or two generations ago. We need to expand our education system and provide the best students with the opportunity to study far beyond their Masters and Doctorates. To do otherwise is to perpetuate the simple scientist, and the silent ecologist.

CONCLUSION

The world needs scientists who are prepared to work — if need be, to fight — for conservation. This means more than carrying out laboratory and field experiments and publishing the results in scientific journals, however worthy those goals may be. It means becoming part of the community, of getting involved with and for people. It means devoting a significant portion of our time and resources to solving environmental problems and communicating with the public (Ehrlich 1993, 1994). It means being an advocate for those who cannot speak for themselves —

future generations and other species. More of us who are concerned about the future and the conservation of Australia's biological resources should enter politics (Recher 1990b). All ecologists concerned for the future should be political and public.

ACKNOWLEDGEMENTS

Judy Recher helped with correcting the manuscript. My ideas have developed in discussions with many people, but especially with W. E. Davis, Jr., Paul and Anne Ehrlich, Dan Lunney, Jonathan Majer, Judy Recher, Paul Recher, and Denis Saunders.

REFERENCES

Davey, S. M., 1989. Thoughts towards a forest wildlife management strategy. *Aust. For.* **52:** 56–67.

Ehrlich, P. R., 1993. Communication: how can ecologists get their message out? Pp. 295–301 *in* Nature Conservation 3: Reconstruction of Fragmented Ecosystems ed by D. A. Saunders, R. J. Hobbs and P. R. Ehrlich. Surrey Beatty & Sons, Chipping Norton.

Ehrlich, P. R., 1994. Enhancing the status of population biology. *Trends Ecol. Evol.* **9:** 157.

Ehrlich, P. R., 1996. Environmental anti-science. *Trends Ecol. Evol.* **11:** 393.

Ehrlich, P. R. and Ehrlich, A. H., 1996. Betrayal of Science and Reason. Island Press, Washington, D.C.

Hall, C. M., 1988. The "worthless lands hypothesis" and Australia's national parks and reserves. Pp. 441–58 *in* Australia's Ever Changing Forests ed by K. J. Frawley and N. Semple. Department of Geography and Oceanography, Australian Defence Force Academy, Campbell, A.C.T.

Kirkpatrick, J., 1996. President's report. *Bull. Ecol. Soc. Aust.* **26:** 38–40.

Lunney, D., 1991. The future of Australia's forest fauna. Pp. 1–24 *in* Conservation of Australia's Forest Fauna ed by D. Lunney. Royal Zoological Society of New South Wales, Mosman.

Martin, B., 1981. The scientific straightjacket: the power structure of science and the suppression of environmental scholarship. *The Ecologist* **1:** 33–43.

Martin, B., 1992. Intellectual suppression: why environmental scientists are afraid to speak out. *Habitat* **20:** 11–14.

Norton, T. and Lindenmayer, D., 1991. Integrated management of forest wildlife: towards a coherent strategy across state borders and land tenures. Pp. 237–44 *in* Conservation of Australia's Forest Fauna ed by D. Lunney. Royal Zoological Society of New South Wales, Mosman.

Recher, H. F., 1990a. Ecology, forest management and the timber industry. *Aust. Biol.* **3:** 85–88.

Recher, H. F., 1990b. Ecology, politics and the environment: Why I stood for the Senate. *Bull. Ecol. Soc. Aust.* **20:** 2–4.

Recher, H. F., 1992. Simple journalists or simple scientists?: Are environmental issues too complex for the media? *Aust. Zool.* **28:** 19–23.

Recher, H. F., 1994. Why conservation biology? Pp. 1–15 *in* Conservation Biology in Australia and Oceania ed by C. Moritz and J. Kikkawa. Surrey Beatty & Sons, Chipping Norton.

Recher, H. F., Shields, J., Kavanagh, R. and Webb, G., 1987. Retaining remnant mature forest for nature conservation at Eden, New South Wales. Pp. 177–94 *in* Nature Conservation: The Role of Remnants of Native Vegetation ed by D. A. Saunders, G. Arnold, A. Burbidge and A. Hopkins. Surrey Beatty & Sons, Chipping Norton.

Resource Assessment Commission, 1991. Forest and Timber Inquiry Draft Report, Vol. 1 and 2. Australian Government Publishing Service, Canberra.

Scotts, D. J., 1991. Old-growth forests: their ecological characteristics and value to forest-dependent vertebrate fauna of south-east Australia. Pp. 147–60 *in* Conservation of Australia's Forest Fauna ed by D. Lunney. Royal Zoological Society of New South Wales, Mosman.

Seager, J., 1994. Science and environmental research: a feminist critique. Pp. 107–15 *in* Restoring the Land. Environmental Values, Knowledge and Action ed by L. Cosgrove, D. Evans and D. Yencken. Melbourne University Press, Melbourne.

Shaw, W. W., 1983. Integrating wildlife conservation and timber production objectives in Australian forests. *Aust. For.* **46:** 132–35.

State of the Environment Advisory Council, 1996. Australia: State of the Environment 1996. CSIRO Publishing, Collingwood, Victoria.

CHAPTER FIVE

Can science meet bureaucracy?

PETER BRIDGEWATER[1] and TONY PRESS[1]

Scientists and bureaucrats have locked horns in recent years directly as a result of changes in the way the governments and the community see both science and government. Bureaucrats have been asked to "do more with less", which in the field of science translates to contracting out research in a competitive market. Government scientific organizations have been partly privatised or corporatised, and formidable citadels such as CSIRO have been required to "find external funds". In this rapidly changing milieu a great deal of misunderstanding has developed concerning the nature and functions of institutions. We argue that science can meet bureaucracy but romantic dreams of the good old days will not help. Critical evaluation of the role of bureaucrats and scientists is the only way to ensure that public and government expectations can be met into the future.

INTRODUCTION

GOOD bureaucratic outcomes for the environment must have a sound scientific base. Of course, a bureaucracy lacking any scientific basis can survive for quite a while and, of course again, good science can function without bureaucratic support, but science cannot function effectively without funding and political support.

Bureaucracy is best defined as effective public administration — the "action layer" between community expectation and demand and political policy and decision making. We should note that although this paper uses the word *bureaucracy* in the title, it is best expressed as *public administration*, and, as such, is part of a continuum between science and technical advice and the political process. There are interactions between all of these three elements. The process of public policy formulation is definitely non-linear!

As scientists within public administration we have a reasonably clear idea of what public administration needs from scientists, especially ecologists, in assisting governments to formulate or implement policies and programmes. In summary, the key needs are:

- objective, comprehensive and defensible data;
- a clear understanding of the limitations of those data; and
- interpretation of those data given specified parameters and contingencies.

The ability to rely on objective and robust data is a far more powerful tool of public policy making than reliance on competing opinions of "experts in the field". Sound public policy making is more likely to evolve from solid objective data and its interpretation, than from simple speculation.

Some years ago the UK Government started demanding their research units "earn" some part of their funding from outside sources. There was then a tendency for the environmental research unit to become involved in projects with development, rather than conservation, as the main objective. Such projects often resulted in the production of an Environmental Impact Assessment.

On the other side, the agency concerned with nature conservation tended to focus on a conservative approach to development questions, or to react negatively to Environmental Impact Assessments. This was especially so if the matter in question involved the real or perceived threat to populations or species which had statutory implications for the conservation agency or its Minister.

What started happening frequently, of course, was government scientists — from the research unit and the conservation agency — appearing on opposite sides and clashing in inquiries and judicial appeals. The adversarial approach used in inquiries does force witnesses into rigid answers although the question may have areas of uncertainty or ambiguity. Such clashes became confusing for the public, irritating for the government, and sometimes devastating for individual scientists involved. The administrative layer was often left to clear up the resulting unsatisfactory situation, often without the benefit of clearly expressed scientific views — even if those views might contain uncertainty or ambiguity.

[1]Department of Environment, Canberra, Australian Capital Territory, Australia.

Pp. 16–19 *in* ECOLOGY FOR EVERYONE: COMMUNICATING ECOLOGY TO SCIENTISTS, THE PUBLIC AND THE POLITICIANS ed by R. T. Wills and R. J. Hobbs. Surrey Beatty & Sons, 1998.

SOME EXAMPLES OF INTERACTIONS BETWEEN SCIENCE AND BUREAUCRACY

Protected Area Management

Protected Area Management is an excellent example, in any country, of interaction (or not) between science and bureaucracy. Kakadu National Park, and Uluru — Kata Tjuta National Park, are both managed by Environment Australia. They both have a formal process of managing scientific knowledge in a bureaucratic and political framework through respective Plans of Management. The management of these Parks is driven by the need to have safe and attractive tourism and, at the same time, to achieve a high level of nature conservation. A further complexity is the need, for both scientific and practical outcomes, to ensure cultural values and Aboriginal scientific knowledge are at the core of the management regime.

At first blush, Plans of Management for National Parks are classic bureaucratic documents, itemising, as they do, a series of actions and objectives for a set time period. Within their administrative rubric, the Plans of Management are also documents which detail a rigorous scientific management system. The community at large (and the scientific community in particular) can use them thus to evaluate management performance. In the case of Environment Australia, the Plans of Management are open for public comment at two points in their genesis, and thus have the benefit of input from scientists and non-scientists. They are also, therefore "peer-reviewed" — an important aspect in ensuring transparency and objectivity.

The legislative regime which gives bureaucratic life to this planning process provides for consideration of scientific and non-scientific aspects of park management, especially the natural and cultural environment, as well as for social and economic impacts of management actions.

Thus, science *can* and does meet bureaucracy in this process. The plans themselves become part of the ongoing scientific evaluation of management actions, as well as the resources being managed. Indeed, if management planning for protected areas is not a process well grounded in science, then the plans simply become administrative documents, suitable to guide placement of toilet blocks, but not to help in ecosystem management.

Whaling policy

Internationally, whaling is a classic example of the interaction between science and the international bureaucracy. Firstly, the International Whaling Commission (IWC) spent many years during the 1950s and 1960s supervising the culling of large numbers of whales. It was almost a systematic species approach, with the largest whales (Blue) being driven to near-extinction first, then smaller whales (Fin, Sei, Right and Humpback) similarly so. In all this relentless exploitation, the IWC maintained a commitment to science, with its Scientific Committee performing regular counts of whales, developing population models and, later in the piece, developing the so-called *New Management Procedure,* a population model designed to ensure population crashes did not happen, and that catch per unit effort remained economic. Thus quotas for culling of particular whale species were systematically introduced. Those quotas, which were derived from the models, were all designed to operate on the basis of a scientific input.

Despite the apparent integrity of the scientific data gathered, and the population models used, whale species continued to decline dramatically. In 1982, the sharp decline in observable population numbers forced the IWC to adopt a moratorium on whaling. The moratorium took over four more years to become reality, with, even then, some nations taking formal reservations on (i.e., not agreeing to abide by) the proposal. In the end, the paucity of whales available to catch underlined the scientific message and forced the issue.

Since that time there has been effectively no commercial whaling, although Norway announced a very limited resumption in 1992, and announced a larger quota for 1996. Since 1986 the scientific committee has been considering what the levels of populations the various whale species were, together with a detailed re-appraisal of the management procedure. This re-appraisal, still to be finalised, will result in a new model known as the *Revised Management System*.

In 1994 (Donovan 1996), evidence was produced that the then USSR had, over a period of years, significantly under-reported catches of the larger whale species, preferring instead to suggest they were catching other species. All these data fed into the population models designed to produce quotas, described. The reason for failure of those models is now clear — it was not so much the models themselves, but the adequacy of data fed into them. Sound like a lecture from "*Introduction to Science 101*"?

Despite all this endeavour, the political will to resume whaling, or be seen to embrace any step on the road to a resumption, is quite absent amongst most members of the IWC. Even if the data have been less than perfect, the science is good, and will not "spoil", even if the political will to use it in a "use" context is not there. In any event, for sound conservation planning we need science to keep a clear understanding of what is happening to whale populations, and we need science to set whale population dynamics in a context of environmental dynamics.

Forestry policy

Perhaps the most current example of the interaction of science, the bureaucracy and the political process are the developments resulting from Australia's National Forest Policy Statement. This statement, which has been agreed by all States, Territories and the Commonwealth charts "... the objectives and policies for the future of Australia's public and private forests" (Commonwealth 1992). Framed in the context of ecologically sustainable development, the policy sets out a number of goals, the achievement of which is founded on the application of the best available scientific knowledge.

The goals of the policy include:

- the maintenance of an extensive and permanent native forest estate in Australia;
- the management of that estate in an ecologically sustainable manner;
- the conservation of biological and the protection of heritage, Aboriginal and other cultural values.

On the production side of forests management, the policy aims to develop an internationally competitive, and ecologically sustainable industry. From a conservation perspective, the policy aims to integrate *in situ* conservation with ecologically sustainable forest management on both private and public lands outside the reserve system.

The *National Forest Policy Statement* evolved from major reports to Government — including the *Resource Assessment Commission Forest and Timber Inquiry* and the report of the *Ecologically Sustainable Development Working Group on Forest Use*. These reports had significant scientific input, although the *National Forest Policy Statement* itself is the product of significant bureaucratic and political process.

The reports mentioned above contributed to the debate on forest use, but scarcely modified it. After many years of acrimony, public campaigning and fractured policy making the National Forest Policy Strategy is now attempting to construct an agreed, scientifically based frame-work for the future of forest management in Australia.

The Policy aims to integrate conservation with forest use by continuing to assign areas to a comprehensive, adequate and representative reserve system, and by complementary management outside reserves in both the private and public forest estate.

The establishment of the "comprehensive, adequate and representative" system of forest reserves throughout Australia's native forests is the subject of ongoing scientific investigation, discussion and debate (see Pressey, this volume). The reserve design criteria are being formulated to encompass biodiversity conservation, endangered and threatened species and communities, "old growth", wilderness and heritage criteria. Technical (i.e., scientific) consideration of reserve design, appropriate boundaries, genetic diversity and replication are integrated into this process. The strength (or weakness) of the outcomes of the National Forest Policy situation will depend on the credibility of the process, both in design and in application. Here we have a classic example of interactivity between the bureaucratic, political and scientific process: on the one hand the bureaucrats negotiating the outcomes of the forest policy are basing the strengths of their negotiations on the veracity of the scientific input into the process; on the other hand scientists are relying on the politicians and bureaucrats to deliver the best possible outcomes from the process.

WHAT LESSONS CAN BE LEARNT FROM ALL THIS?

Firstly, science is vital to the management of natural resources, whether for conservation or utilization. Its importance is both at the decision-making end: what, where, how and how much; and at the process end: monitoring, impact evaluation; adaptive management and rehabilitation and restoration.

Secondly, you can have a perfect scientific basis for decision making, but if the bureaucratic shell is inadequate, or the political reality is that another set of decisions will influence the outcome, then science can lose its primacy. This does not diminish the value of science, but it does affect its context. We note, with interest, the debates in recent years among scientists about making their work more "relevant". One response to this has been participation in the political debate. For some this path has been fatal, largely because good scientists do not always make good advocates. We suggest, therefore that a keen knowledge of the ecology of the decision making process is as critical as the ecological issues themselves!

In the case of the IWC there has been the resignation of a Chair of the scientific Committee, ostensibly on the basis the IWC was neglecting science. Well, it may have been, but science will always be subservient to the needs and aspirations of society, and one of the skills scientists need is to keep their science fresh and relevant. Scientists need high-wire skills to keep their views and information to the fore, without compromising, or being compromised by, the (always more overt) needs of the political process.

A real problem is that good ecological data are more frequently found in the "grey" literature than in journals. These data can reside in personal, private or government data bases, in filing cabinets, old notebooks and locked drawers. Often crucial observational data can exist in the experience of an individual. Open public processes in high profile decision making is one way to unlock some of these data, but accessibility and comprehensiveness are two problems that will always be present in this kind of process. The issue of dealing with "grey literature" is a major one, as it is often contested as lacking credibility due to the absence of peer review.

How do we deal with these situations, especially when we rarely have a complete data sets, gaping holes in research effort, and limited time and resources to deal with complex ecological questions in the time frame of government decision making? Should we be more honest, and say that we can only do our best with the available data? In addition, there is an expectation that we will apply the "precautionary principle". Application of this principle is especially difficult when government decision timetables world-wide are forcing bureaucrats to make crucial decisions on limited data sets — if the precautionary principle were literally applied nothing would ever happen!

It will surprise no-one that we have no stock answers to these conundra, but they represent part of the ongoing debate and discussion that has to occur between scientists and bureaucrats — and that means scientists and the wider community. To be effective, scientists must not "talk up" possible solutions or options — there is rarely "one" or "a right" way to solve a problem. And bureaucrats must accept and understand that scientists cannot always produce results to order — or that, if they do, then all participants are involved in an experiment. Which, to paraphrase former Prime Minister Margaret Thatcher, brings us to climate change . . . !

REFERENCES

Commonwealth of Australia, 1992. National Forest Policy Statement. Canberra.

Donovan, G. P. (ed.), 1996. Forty-Sixth Report of the International Whaling Commission. International Whaling Commission: Cambridge, UK. 688 pp.

CHAPTER SIX

Ecologists in public

RICHARD J. HOBBS[1]

"The object of education lies not in communicating the values of the past, but in creating the values of the future" —John Dewey

Traditionally, ecologists have focused on scientific questions relating to the interactions between organisms and their environment. However, it is becoming increasingly important for ecologists to develop skills beyond the traditional scientific training. Important among these are the skills necessary to communicate well with a range of audiences. Communication is necessary between scientists and other scientists, both within and between disciplines, between scientists and managers and policy makers, and with students and the general public. Skills for increasing the effectiveness of these different types of communication can be learned and need to be included in scientific training. More effective communication can only assist in the process of ensuring that science is perceived as a useful and necessary activity and that scientific results reach those who need to use them.

INTRODUCTION

WE live in times of rapid change and uncertainty about the future. This change and uncertainty has many facets, including political, social and environmental, and affects everyone individually and collectively. In this chapter I explore the implications of this situation for scientists, and in particular, ecologists.

Ecologists traditionally conduct science on how organisms interact with other organisms and the environment. The training an ecologist has had equips him/her to investigate these interactions effectively and rigorously. However, that training frequently does little to equip ecologists for another set of interactions — those that occur between scientists, and between scientists and other parts of society, particularly managers, policy makers and the media. This set of interactions is becoming increasingly important as the need for closer links between science, management and policy is recognized, and as more scientists either expose themselves willingly to the media limelight or are dragged there. Can ecologists improve their ability to influence policy and management? Can they make more efficient use of the media to help make the general population ecologically literate? These questions are not trivial, and the answers are likely to determine the future for ecologists (whether they have jobs or not), for the environmental and economic state of the country (whether it has any topsoil, biodiversity etc.), and ultimately the planet (whether we still have an inhabitable one or not).

I argue that a radical shift in the behaviour of scientists is called for, which would see scientists more actively engaged in communication activities on a wide front. I first explore the reasons why scientists need to communicate and who they need to communicate with. I then discuss the additions to the scientist's "tool kit" that are required to make successful communication possible, and point out that increasing levels of information supply mean that scientists are, in essence, competing for audience attention. Attracting and reaching an audience is thus a vital part of the process, and I outline relevant ideas concerning communication and learning methods.

WHY COMMUNICATE? WITH WHOM?

If one stops to look at the calls for scientists to communicate more, a common reason for advocating this is that scientists need to inform people more about what they are doing and why it is important. Why inform people, though? Communication should not be about simply informing people but more about motivating them to change their actions or beliefs. These changes may take different forms, depending on the audience.

[1]CSIRO, Division of Wildlife & Ecology, Private Bag, P.O. Wembley, Western Australia, Australia 6014.
email richard.hobbs@per.dwe.csiro.au

Pp. 20–25 *in* ECOLOGY FOR EVERYONE: COMMUNICATING ECOLOGY TO SCIENTISTS, THE PUBLIC AND THE POLITICIANS ed by R. T. Wills and R. J. Hobbs. Surrey Beatty & Sons, 1998.

There are a variety of different targets for communication by scientists, which can be summarized as follows:

1. *Fellow scientists (same discipline).* While many scientific endeavours involve individual research, there is still a need to communicate with other scientists in the same field. This takes the form of writing scientific papers and reports, attending and presenting papers at scientific meetings, and the like. While university training provides scientists with the fundamentals needed to perform adequately in their chosen fields, it frequently fails to provide much, if any, guidance in how to write and speak effectively.

   Scientific writing aims to communicate research methods and results in an unambiguous and clear manner, and there are simple guidelines available for achieving this (Lertzman 1995; Lindsay 1995; Magnusson 1996; Matthews *et al.* 1996). Unfortunately, many scientific authors and editors also appear to believe that something is adequately scientific only when it is also unremittingly boring. This need not be the case! While there is undoubtedly a need for clear writing and adherence to grammatical correctness, there is room for creativity and good prose. One needs only to observe the scientific writing styles prevalent earlier in this century to see how much we have removed the poetry and beauty from modern scientific writing. Essay writing is also a declining art, and yet well constructed essays are a powerful motivating and inspirational force.

   Much the same can be said about scientific presentations at meetings. These need to be clear, well structured, well illustrated with uncluttered illustrations, and designed to fit within the allocated time slot. But they also need to be *interesting*. At the average scientific meeting there are usually dozens, if not hundreds, of presentations, often in concurrent sessions. Each presenter is therefore competing, first, for the audience's presence and attention, and second, for a place in their collective memory afterwards. Again, many scientists seem to consider that their subject matter is serious stuff, and that there is no room for humour, passion or originality in their presentation. Certainly, not everyone can be an ace presenter, but there are clear guidelines on how presentations can be made as effective as possible. This is equally true of poster presentations. Given the enthusiasm with which most scientists approach their work, should not this enthusiasm overflow into how they present it?

2. *Other scientists (different disciplines).* Increasingly, the need for greater interactions between disciplines is being recognized, especially in relation to finding solutions to environmental problems. The ecologist will be expected to interact with many other scientists, including hydrologists, soil scientists, agronomists, geologists, atmospheric physicists, to name but a few. Beyond that, the need for closer interactions between the sciences and the social sciences is being recognized, and ecologists may also have to communicate effectively with economists, planners, sociologists and others. Communication involves breaking down old disciplinary rivalries and barriers, achieving a common language, and recognizing the value of different approaches and the need to combine work to deal with the big issues confronting us today. This communication becomes particularly important where interdisciplinary teams come together to tackle particular problems. The success or failure of the team will in large part depend on how effective the communication between team members is.

3. *Students.* Many scientists are based in universities or other teaching institutions, where part of their duties involve teaching students. Here again, while most scientists have a sound training in their subject, they often do not have adequate training in teaching skills. University lecturers are often appointed on the basis of their research record or potential, and rarely on their teaching skills. How many dreadful lectures do you remember sitting through as a student? Such a situation is easily fixed by the adequate allocation of resources to training academic staff in elementary teaching techniques, as already happens at some institutions. More than simply learning new techniques, however, academic staff need to recognize the importance of educating students to think critically. This may be uncomfortable to some, since it entails providing the student with the skills to criticize the instructor. Nevertheless, it is an essential element in the development of science and societal change.

   Scientists also have to communicate with students undertaking research degrees. Good communication between supervisor and student is essential for the successful completion of any Ph.D. or Masters degree, and failure can often be traced back to the breakdown in that communication. Adequate communication requires that the supervisor not only acts as an academic mentor, but is also able to communicate with the student as one human to another. Again, the development in the student of the ability to think critically is an essential element.

4. *Funding bodies* The search for research funding is becoming an increasingly large and time-consuming task, but few would view it as an exercise in communication. However, that is exactly what it is. The researcher is trying to convince the funding body, or the referees appointed by that body, that his/her idea for future research is worth funding. Here again, he/she is competing for the funders attention and dollars with many others, and the merits of the proposed research will be more readily apparent if the proposal is well written and structured.
5. *Policy makers/managers.* If the scientist is content to "do science" and not worry about whether it is used or applied, he/she may never need to talk to a manager or policy maker. However, the days when such ivory-tower behaviour was appropriate have gone, and it is becoming increasingly important that science flows through to management and policy (see Gascoigne and Metcalfe, this volume). This demands meaningful interactions between scientists and "user groups" or "stakeholders". These interactions are still apparently not as frequent as their importance would demand (Wilson and Barnes 1995). Cleaves (1994) suggests that "Society's changing values toward foresters [read "ecologists"] and the billowing complexity of forestry [read "environmental"] issues have created the need for [a] new role — more visible, more exciting, and more dangerous than the traditional one. To play this part well, we will have to learn our lines, study the context, and develop new skills. The payoffs are great. And if we don't at least audition, the policy show will go on without us — and someone else, perhaps someone less qualified will get the part".

   Increased interactions with policy makers and managers require the scientist to be able to communicate clearly and effectively without recourse to what Aldous Huxley has called "specialized meaninglessness", i.e., the jargon in which scientists often immerse themselves. It also requires that they are able to put their work and interests in the correct context and can indicate the relevance of it to those who wish to use it. Increasingly, too, managers need to be involved in the planning and implementation of the research, especially where this involves changes in land use or management practices. The concept of adaptive management is relevant here. Communication with managers and planners during the planning of research projects can assist in assuring that the research is directed in ways that will be most useful, since managers can bring a useful perspective on what options are practically feasible and hence worth pursuing.
6. *"The public".* We frequently hear calls for scientists to be more actively involved in communicating with the public, and engaging in public debate. Sir Michael Atiyah, outgoing president of the Royal Society was recently quoted in New Scientist as saying, "Scientists are too often thought of as a secretive élite, a sinister part of the establishment, part of "them" not "us". The only way to break down this suspicion and distrust is for scientists to speak out openly and freely". Current anti-science and anti-environmental trends can be in part explained by the failure of scientists to communicate effectively with the rest of the population.

   Ecologists potentially have a large role to play in public debate on environmental issues. However, the actual process of getting information to the public is often a bit of a mystery to many scientists. The main ways that it occurs are through scientists' contacts with interest groups, and via the media. All the same considerations apply as were discussed above in relation to communication with other groups, and there are a series of other considerations dealt with by other contributors to this volume. There are, of course questions concerning scientists' willingness and/or ability to speak openly about contentious issues, and progress in this area will come only from a recognition by individuals and organizations that open debate is both desirable and necessary.

   One set of members of the public which bear special attention are the children. They represent the future, and it is our responsibility to ensure that we equip them as well as possible for that future. Communicating effectively with children is a difficult but very rewarding task, which not everyone has the knack or desire to try. Contact with children requires a clarification of ideas and concepts far beyond that usually required in scientific circles, and children have a knack of asking very piercing questions. Nevertheless, even if every scientist is not involved in direct contact with children, they could at least be involved in making information available to those who are. Organizations like the Australian Association for Environmental Education do an excellent job of co-ordinating and initiating education activities.

THE ECOLOGIST'S TOOL BOX

The previous section illustrated that there are many reasons why the scientist should be involved in communication, and many different audiences to communicate with. The traditional tool box of an ecologist has therefore to be expanded to include a new set of skills involving communication (Fig. 1). These skills thus need to be included in the training provided to ecologists, and scientists in general. Some scientists can pick them up "on the run", but that is not a very efficient way of increasing the effectiveness of scientists in general.

There is currently an explosion in information supply due to advances in electronic media and communications, although the amount of that information which is consumed and applied is increasing at a much slower rate (Fig. 2). This information covers all aspects of life including social, political, religious, cultural, sporting and many other issues. This means that the ecologist has to compete effectively with an array of other "information providers" to make the target audience listen and take notice. His/her communication activities must therefore aim to make the maximum impact possible. Part of the toolbox must therefore be effective communication methods. There is

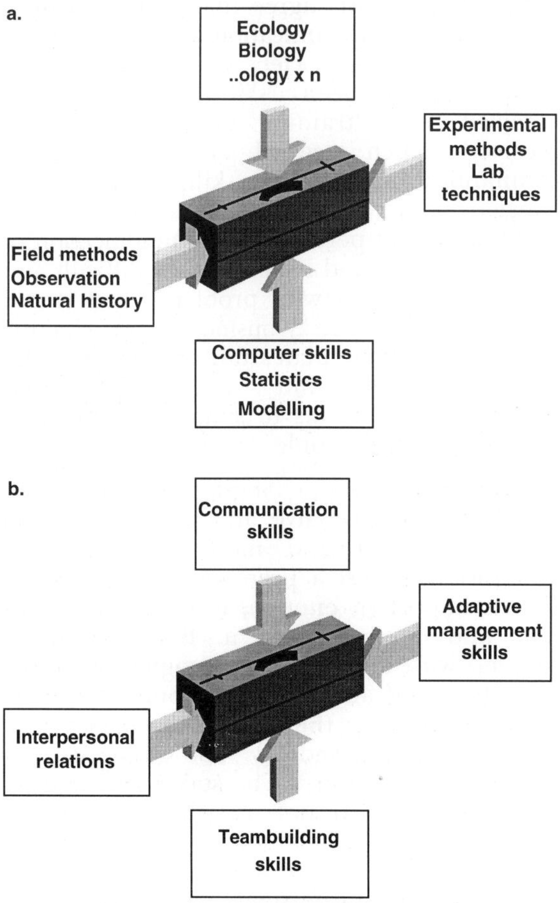

Fig. 1. The traditional ecologist's tool box (a), and the additional components required in today's world (b). (Adapted from Kessler 1995.)

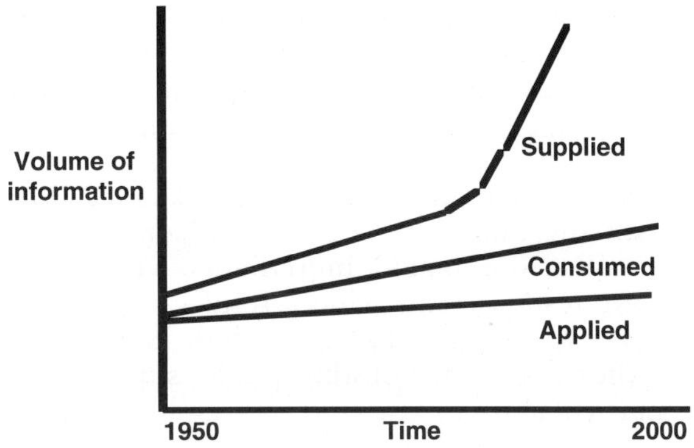

Fig. 2. The amount of information supplied, consumed and used over the past 50 years. (Redrawn from van Cuilenberg 1987.)

considerable information available on communication theories, learning styles and so on (see, e.g., Knowles 1991; McQuail and Windahl 1993), which is directly relevant to this endeavour, but of which most scientists are totally unaware.

While this is not the place to go into detail, it is worth noting that several different styles of learning, or receiving and interpreting information, are recognized, and this can assist in developing communication methods which have the greatest chance of success. It is important to recognize that the people being communicated with come complete with a package of beliefs, values, cultures, and levels of education. These may directly conflict with those of the communicator, or there may be conflicts between different members of the audience. It is always a mistake to believe that the people you are trying to communicate with think and feel the same way you do about particular issues or about life in general. Some recognition and acceptance of differences in opinion and viewpoint makes a much better base for successful communication. In addition to this, ecologists face the problem that different people will interpret the same message in different ways, especially when the message is simplified or presented in non-technical language. Thus the message needs to be presented in a way which will change the way the audience thinks, without antagonizing them in the process.

In addition, different people respond to different stimuli. For instance, some people respond more to visual than to auditory information. Others take in numeric or quantitative data much more easily than qualitative information, and vice versa. Some need concrete data and examples, while others gain more from conceptual information. People respond to humour, gravity and passion in a variety of ways, depending on the situation.

All of these considerations can be used to construct effective communication packages. These can be tailored to the type of audience, or they can be generalized to try to hit as many targets as possible. Frequently it is not the message itself, but how it is packaged, that makes the difference. Passion and enthusiasm are often the magic ingredients which will inspire response and achieve results — whether this is expanding a student's mind, influencing a policy maker or enlisting public support for a particular issue.

Successful communication

The important ingredients needed for the successful communication of issues in a forestry context were recently summarized by Cleaves (1993). I suggest that they have broader applicability to anyone interested in natural resources, and provide a useful framework for effective communication. I provide an abbreviated list of Cleave's main points below. While these deal primarily with speaking to groups, the principles are more generally applicable to any form of communication.

1. Understand the issues — i.e., know what you are talking about so that you do not come over as a half-wit. Know your facts, make sure your data are correct, and be prepared for questions from the left field.
2. Realize how opinions are shaped — the points discussed earlier. Know how to influence people's opinions and perceptions.
3. Know your listeners — figure out what sort of audience you will be dealing with and tailor your communication efforts accordingly.
4. Be specific — generalities are rife in science, but are not much use in real life situations.
5. Have a clear purpose and strategy — know what you want to communicate and why.
6. Be calm — adequate preparation is always a prerequisite.
7. Do not blame — apportioning blame immediately puts a proportion of your audience off side. If blame is appropriate, let others figure it out for themselves.
8. Focus on the facts, but identify opinions and values — scientists are supposed to deal in facts, but it is important to give informed opinions too. It is also important to admit when you do not know the answer to something. Do not provide spurious certainty where uncertainty is appropriate.
9. Raise questions — "experts" know better than anyone exactly how much we do not know. Point out unknowns and indicate where more research is appropriate (without sounding like an appealathon for research funds).
10. Be brief — waffle and obfuscation lead to bored and disinterested audiences. In particular, do not exceed your allotted time — it annoys everyone and indicates an arrogant disregard for other people. Also, allow time for questions
11. Practice — the only way to become good at anything is to practice. The more you do, the better you become.
12. Follow up — part of effective communication is persistence. Where appropriate, show your audience you are interested and concerned by continued contact
13. Keep at it — things rarely change overnight. In fact, it often takes concerted effort over months or years to achieve anything.

CONCLUSION

Not all ecologists agree that there is a need for enhanced communication between scientists and other parts of society. Communication has traditionally not been a skill in which scientists received much training or for which they received much reward. Indeed, effective communication requires skills that can be learned, but beyond that also demands enthusiasm and persistence. Communication of ecological ideas and issues is often difficult and potentially fraught with problems, and is still seen by many as a risky business. Most scientists already have their plates full and are juggling numerous tasks and activities, so that communication seems to simply add more to an already full schedule.

Nevertheless, the changing nature of the world in which we live and work makes the need for better and more effective communication almost a prerequisite for survival. If it is important enough, ecologists will find time and energy to do it. Better training, coupled with an increased recognition of the importance of all aspects of communication, can only improve the science of ecology and how it is perceived by non-ecologists. As Daie (1996) suggests, "The stakes are high. If not now, when? If not us, who? Clearly the responsibility is ours alone. If necessity is the mother of invention, and if scientists are unresponsive to professional evolution, then is it not time to invent a new species of scientist?" When the winds of change are blowing around us, we can elect to build either a shelter or a windmill. Effective communication can help us harness the forces affecting us rather than letting them blow us away.

ACKNOWLEDGEMENTS

I thank Bert Main and Denis Saunders for insightful comments on the draft manuscript, Monica Martinovich for the inspiration to explore communication issues in more depth and my children, Katie and Hamish, for the realization that we do not need big words to communicate the wonders of the world to others.

REFERENCES

Cleaves, D., 1993. Speaking as a professional. *J. For.* **91**(7): 25.

Cleaves, D. A., 1994. Foresters as issue educators. Working at the roots of policy. *J. For.* **92**(3): 8–12.

Daie, J., 1996. The activist scientist. *Science* **272:** 1081.

Kessler, W. B., 1995. Wanted: a new generation of environmental problem solvers. *Wildl. Soc. Bull.* **23:** 594–99.

Knowles, M., 1991. The Modern Practice of Adult Education. Cambridge University Press, Cambridge.

Lertzman, K., 1995. Notes on writing papers and theses. *Bull. Ecol. Soc. Amer.* **76:** 86–90.

Lindsay, D., 1995. A Guide to Scientific Writing (2nd ed.). Longman, Melbourne.

Magnusson, W. E., 1996. How to write backwards. *Bull. Ecol. Soc. Amer.* **77:** 88.

Matthews, J. R., Bowen, J. M. and Matthews, R. W., 1996. Successful Scientific Writing: a Step-by-Step Guide for Biomedical Scientists. Cambridge University Press, Cambridge.

McQuail, D. and Windahl, S., 1993. Communication Models for the Study of Mass Communication. Longman, London.

van Cuilenberg, J. J., 1987. The information society: some trends and implications. *Euro. J. Com.* **2:** 105–21.

Wilson, S. and Barnes, I., 1995. Scientists' participation in environmental policy. *Search* **26:** 270–73.

CHAPTER SEVEN

The importance of networks for communication and application of ecological research

DENIS A. SAUNDERS[1]

In designing and conducting research, ecologists and conservation biologists should ask themselves: why and for whom they are conducting the research; what can be done with the results; and how they can disseminate the results to influence environmental management? They should devote more attention to developing and maintaining appropriate networks to enable them to ensure that the results of their research are used to improve the management of our environment. Such networks may involve members of the community including school groups, family groups, local government officials and members of landcare groups, as well as local and international volunteers, other scientists, members of the mass media and politicians. Such networks can be influential in decisions made on matters relating to the environment.

INTRODUCTION

WE live in a time of rapid change and nowhere is this more obvious than when we look at changes in the natural environment. I started work as an animal ecologist with CSIRO, an Australian Government research establishment, in January 1968. Since then I have spent my time working on avian species and communities in agricultural landscapes and on island ecosystems. Unfortunately, all of my studies reflect this rapid change and much of my work has involved cataloguing the decline of species and communities. For example, my first study, and one in which I am still involved, was on the ecology of Carnaby's Cockatoo *Calyptorhynchus latirostris* which has disappeared from more than a third of its breeding range in southwestern Australia over the past thirty years because it is unable to cope with the changes imposed on its landscape; unfortunately its range is still contracting (Saunders 1990).

My most recent work has not been research, but has been picking the eyes out of other peoples' research and writing by committee. From August 1994 to July 1996 I chaired the reference group which drew up the chapter on biodiversity for the 1996 Australian State of the Environment Report (Commonwealth of Australia 1996). The main findings of the biodiversity chapter reflect the rapid and, in many cases, detrimental environmental changes taking place:

> *For those terrestrial groups for which there is sufficient information to assess current state, the trends are alarming. Five per cent of Australia's higher plants, 9% of birds, 23% of marsupials, 7% of reptiles, 16% of amphibians and 9% of freshwater fish are vulnerable, threatened or extinct. These downward trends are continuing* (Commonwealth of Australia 1996).

The introduction to the Australian State of the Environment Report states that many of the environmental problems facing Australia have solutions, however, most are not being put into practice. The report notes that the overarching threat to biological diversity is posed by the numbers of people, their life styles, and their demand for natural resources, including those for export. It also reports that removal of native vegetation and the flow-on effects of fragmentation of habitat and changed ecological processes are the major causes of loss of Australia's unique and incredibly rich biological diversity (Commonwealth of Australia 1996). Despite the fact that this is well known, we collectively still allow around 500 000 ha of native vegetation to be cleared every year in Australia (Commonwealth of Australia 1995). It seems hypocritical that Australians object to the scale of land clearing in the Amazon Basin, yet they do not make much more effort to change the situation in Australia. Why should we allow more land to be cleared when we patently cannot manage much of the land we have already cleared? At present, there are no adequate assessments of the amount of land being cleared annually, despite the technology being available and some Australian Federal, State and Territory governments saying that such assessments should take place.

While some of the fault for Australia's failure to implement some remedial environmental

[1]CSIRO Wildlife and Ecology, P.O. Box 84, Lyneham, Australian Capital Territory, Australia 2602.

Pp. 26–32 *in* ECOLOGY FOR EVERYONE: COMMUNICATING ECOLOGY TO SCIENTISTS, THE PUBLIC AND THE POLITICIANS ed by R. T. Wills and R. J. Hobbs. Surrey Beatty & Sons, 1998.

action is the result of the lack of commitment and the will to tackle the big picture by politicians and the community (Saunders 1996) some of the fault also lies with ecologists and conservation biologists. They are the scientists with much of the information relating to environmental change and assessment, yet not enough of this information is getting into the public arena to try to change the attitudes of the public to protect the environment better (see Hobbs, this volume). Nor is much of the research being integrated with the results of other studies to produce landscape scale solutions for application. Ecologists and conservation biologists need to lift their heads from the narrow studies they are undertaking, look at the practical implications and applications of their research and findings, and put them into a broader environmental framework, rather than a narrow more restrictive view they often have.

Most scientists operate under the publish or perish management regime; the system that maintains that the work has not been done until it has been published. I believe that, given the serious nature of the environmental problems we are faced with and the fact that ecologists and conservation biologists have a pivotal role in developing an understanding of these problems and possible solutions, they should operate under the adage that:

> the research has not been done until it has been published in a form that can be used by those who are in a position to apply the results to guide their actions to manage natural resources better.

Further to this adage, ecologists and conservation biologists should ask themselves several questions when designing and conducting research:

- Why is the research being conducted?
- For whom is the research being conducted?
- What can be done with the results?
- How can the results be disseminated to influence what happens on the ground?

This does not mean trivialising the research, but rather, after carrying out top quality research, scientists should go that bit further, interpret the results and produce them in a way that they can be seen and applied. Publishing the work in the *Australian Journal of Ecology, Conservation Biology, Landscape and Urban Planning, Pacific Conservation Biology* or other international journals and then expecting the results to be picked up and incorporated into management practice is an extremely naive view of the real world. Few of the decision makers or those managing natural resources have the time to read the scientific literature and then try to work out how to apply what they read to their decisions or management. The work is much more likely to be applied if it is produced in a forum where it is readily accessible and in a form that can be more easily assimilated.

There are many avenues open to scientists to help spread environmental messages and aid the integration of research results to affect use, rather than abuse, of natural resources. Many of these involve the development and maintenance of networks with many individuals and groups, including, but not necessarily in any ranking of importance, members of the public, students, volunteers, other scientists, editors, journalists, politicians, land managers and other people who make decisions on the way natural resources are used. Development of networks with some or all of these groups of people is important and should be a priority for most if not all ecologists or conservation biologists (see Saunders *et al.* 1995) to increase the application of sound ecological practices to environmental management.

INVOLVING PEOPLE IN RESEARCH

Most ecologists and conservation biologists are enthusiastic about their research and enjoy the field work associated with it. This enthusiasm can be harnessed by inviting others to participate in the work or by publicising the work through the community in which the research is being conducted. There are many opportunities to involve members of the community in ecological research. For example, when starting field work in a new area, or when initiating research in an old study area, one could arrange to go to the local schools (primary and secondary), talk to the pupils and explain what is being done, why it is being done and what the environmental implications of the research are. This is often a good way of reaching parents, particularly if the pupils are enthused by the presentation. It can be useful to extend an invitation to the pupils via the teachers for them to visit the field site and take part in the research. CSIRO Wildlife and Ecology has found this a very useful exercise at their study area around Kellerberrin in the central wheatbelt of Western Australia.

Visits by school groups can be very popular and have been the basis of some important and productive networks at Kellerberrin. For example, in the work on the movements of birds in the extensively cleared landscape of the central wheatbelt of Western Australia, a number of school groups have spent part of their school day with us. Some of those students have enjoyed their time so much that their parents have contacted us subsequently and asked to come out and see what the research is about. In one case, a student first came out with us when he was mid-way through high school, and subsequently he came out with us as often as he could. He went on to do environmental science at university and his experiences helping with field work were an important factor in making his decision regarding tertiary education. His father was the head of the major landcare group in the area and he has been supportive of the CSIRO research. He has fostered requests for funding for us and has been keen on receiving practical results his landcare group can apply to address problems of landscape reconstruction. In another case, the whole family took part in field work several times and the father admitted that seeing birds he had not seen before opened his eyes to other elements of the landscape. He has since become a leader in his local sub-catchment group and has a close involvement with the CSIRO research unit. His sub-catchment group is using results from the research to guide their revegetation, with nature conservation being one of the objectives. He has also taken a major leadership role in the new integrated catchment approach being taken in Western Australia and has actively sought the involvement of CSIRO in integrated catchment management. He has also drawn the attention of others to the work of CSIRO Wildlife and Ecology in landscape planning and management and even lobbied the Western Australian Minister of Agriculture to strengthen collaboration between the CSIRO unit and staff from Agriculture Western Australia.

Other community groups, such as landcare groups, shire councils and natural history groups are useful groups to contact and brief on research in their area. In the Kellerberrin area, CSIRO has found the local councils and individual councilors have been supportive of the research and interested in the results. One local councilor assisted a Ph.D. student interested in the ecology of the dominant woodland species by sharing his extensive knowledge of the species, including showing the student areas where the species had regenerated under a range of disturbance regimes.

Local radio stations and newspapers are also very useful for developing contacts with members of the community. Radio stations have a considerable amount of broadcast time to fill and are especially interested in items which effect the local community. For example, when setting up a new study it can be useful to ring the local station and speak to one of the producers. In many cases they will be interested in having the research worker on air to discuss the research. This provides a useful opportunity to seek support and explain the environmental issues.

The involvement of volunteers in field work can be a successful method of developing networks which can have benefits for the research worker, the volunteers and the environment. For example, I have had a long and fruitful association with EARTHWATCH who have supported my study on bird movements (Gilmour and Saunders 1995). Over a period from 1988 to 1995 they provided 152 volunteers from six countries and funds for field work; this covered the labour component and field expenses for 20 two-week expeditions allowing about 420 banding sessions to be worked. The benefits for the scientist in receiving labour and funds to cover the field operations are obvious, however, in the long term the opportunities to educate many people on environmental matters and develop an international network may be much more important for the environment.

People coming from other countries to work in an area often attract the interest of local members of the community and others, who then want to find out why people want to come to work on environmental problems in another country. This provides opportunities to discuss environmental problems, raise environmental awareness, explain the need to modify our demands on the environment and put forward potential solutions. In addition, this "exotic" involvement is often the "new slant" the media wants to run with environmental stories they think may be "old hat".

Eco-tourism is an increasingly popular activity (see van Oosterzee and Preece, this volume). Some field-based ecological research has the capacity to link with tourist operators who wish to bring clients to take part in, or observe, the research. Such ventures may be brief and part of a range of activities, or may take longer and involve one research project. In any event, these activities provide further opportunities to raise environmental awareness. One tour operator spent several days at Kellerberrin with the CSIRO project. As a result, one of the people on the tour was able to discuss the management of road verge vegetation. She was a shire councilor and used her experiences and the links she developed with CSIRO research staff to help in developing a management plan for roadside vegetation in her shire.

Over the course of the field work there are many opportunities to raise the environmental awareness of volunteers, who take this awareness back with them. I have no idea how far the environmental message travels from volunteers to friends, families, work mates, etc., however, there is a set of linkages along which environmental education flows; of course this is not a one way information flow.

An ecological experience can have profound effects on some people. For example, one Earthwatch volunteer, who was an electrical engineer in the defence industry in Washington DC, worked in the bird and kangaroo movement studies with CSIRO in 1988. He was so excited by the research he went back to university part-time to do biology. He came back to work with the programme again and then enrolled full-time in a Ph.D. programme in the USA and has now finished his thesis on the genetics of isolated populations and is working in conservation biology.

Between 1987 and 1990 I conducted a survey of the status of the birds of the wheatbelt of Western Australia with 185 people scattered through the region. They kept records of the species that occurred in their area during the period of the survey. The work has since been published as an atlas of changes in distribution and abundance of the wheatbelt avifauna (Saunders and Ingram 1995). When I wrote the first draft I had it read critically by several people, including one farmer who collected data for the survey and who has probably forgotten more about birds than I will ever learn. He was very helpful and told me that I needed to provide a section which explains what people can do to address some of the problems of our declining avifauna. This I did and I have been told that it has been useful to farmers wishing to plan for conservation as part of their farm planning. I subsequently received strong criticism from several scientists for using everyone who volunteered to assist because they were all not qualified ornithologists and therefore their records were suspect. My counter to this is that if we want to get the community involved in conservation we must encourage them to do so and accept that the data may not be as rigorous as that which may come from so-called professionals. However, if we wait until we have enough professionals to conduct a survey such as the one we carried out, it would never have been conducted.

A productive network developed from the involvement of these volunteers in the survey. Several excellent bird watchers have written, or are now writing, their observations with a view to publishing them somewhere. These are based on notes that go back in some cases over 60 years. They will be invaluable references. At present, I am trying to organize one fascinating example from the northern wheatbelt and the least we will produce is a number of bound photocopies which will be lodged in major research libraries in Western Australia.

My contact with volunteers who have assisted in my research has lead to many opportunities to take part in seminars and field days at which it has been possible to present results and offer possible solutions to some environmental problems. I also found out that while collecting data many volunteers broadened their view of their landscape, finding birds they did not know were there before they started keeping records. For example, one woman wrote to me at the end of the survey that she was sorry that I no longer required her help. She thought she was observant when she started collecting data, however, she found out how much she needed to open her eyes. Her husband was a major force in farm politics and several of her friends were also collecting data for the bird survey. By comparing her bird list with those of her friends she was able to see how certain birds needed particular habitat types and that reinforced the need to protect remnant vegetation. She actually had her husband stop using chemical sprays because she was worried about their impact on invertebrates and through them, on birds like the Grey Shrike-thrush which she had come to treasure.

Although difficult to quantify, there are many benefits from involving others in research, however brief that involvement may be. The networks that may develop from this involvement may in the long run be far more productive for the environment than the research itself.

POOLING EXPERIENCES AND RESEARCH RESULTS

Much ecological research is focused and narrow in approach. This in not a criticism; rather it is a statement of fact. Ecologists and conservation biologists often tend to carry out detailed research on a narrow subject. However, if this research is to have application it often needs to be put in to context with the research of others. This provides the opportunity to develop networks for integration and application. For example, throughout scientific and farming communities there is a great deal known about the way landscapes function and the adverse effects farming systems may be having. However, in order to make the most of the information available for management purposes, it needs to be synthesized, integrated with other relevant research and interpreted in a way that those who are in a position to use the information may do so effectively. It also needs to be published in a form and place where non-scientists will see it.

An example of this approach is shown by a three-day residential workshop in 1990 which the CSIRO Wildlife and Ecology organized for 21 people including agricultural economists and scientists, botanists, farmers, hydrologists, soil scientists, and zoologists to combine their knowledge into a conceptual framework of how to reintegrate the fragmented landscape of the central wheatbelt of Western Australia, taking into account agricultural production and nature conservation. At the outset, the organizers specified two products from the workshop. The first was to be a scientific review of the state of knowledge regarding the central wheatbelt and its problems, together with an outline of a conceptual framework to address these problems (Hobbs and Saunders 1993). This review was published by Springer-Verlag in New York and is regarded by some farmers as essential reading for anyone involved in management of the wheatbelt (Doley 1995). However, it is expensive (about $Aud150) and is not readily available to landcare groups and farmers.

The second product specified was to be a "user friendly guide" to the landscape of the central wheatbelt with suggestions as to how to address some of the major problems (Lefroy *et al.* 1991). This product was the *Revegetation Guide to the Central Wheatbelt,* a 106-page book developed by many involved in the workshop as well as a number of farmers. It describes the wheatbelt catena, the soil types are illustrated in terms of the vegetation that originally grew on them, it provides descriptions of 82 species of local plants available for revegetation, gives information on approaches to revegetation including planning, paying attention to placing species on appropriate soil types, and provides contacts for specific information. The guide specifies local species for revegetation on the basis that they are better suited to the environment and should be of more benefit for fauna. The guide was distributed free to about 3 500 farmers in the central wheatbelt. A survey among farmers conducted to establish what use has been made of the publication and whether it is influencing revegetation (Atkins and Hobbs 1995) revealed that about 30% of the farmers surveyed had read or used it and 6% had changed some aspect of tree planting or revegetation strategies on their farms as a result.

There was a third product of the workshop and one which the organizers did not plan for. This product has been the establishment of a co-operative network linking people from a range of backgrounds and organizations. This has resulted in a number of collaborative ventures relating to landscape reconstruction. One important outcome has been that agricultural scientists are now bringing nature conservation into catchment planning as an important landscape element.

THE CONFERENCE NETWORK

Conferences provide important opportunities to develop networks to pass on and receive information and to assist in the formation of collaborative ventures to integrate research results to tackle environmental problems. Conference organizers often concentrate on the information flow by trying to maximize the number of formal talks, even to the extent of using all available time for formal presentations, while not providing ample opportunities to foster the development of networks amongst the conference participants.

Conferences on specific themes can be powerful ways to focus attention on particular environmental problems, provided the opportunities are available to get the participants mixing to explore ways to integrate their results with those of others to deal with the broader issues. The formal presentations with usually limited time for questions, provide the basis for starting the process, but only if there are opportunities to get together more informally to develop the necessary networks. These opportunities can be provided fairly simply in several ways. For example, holding conferences in smaller towns or cities creates a residential atmosphere which fosters interaction between participants, while holding conferences in capital cities often leads to participants being widely dispersed and limits interactions in the evenings. Arranging for poster sessions to be in the same room as the catering so that people can take sustenance and meet and explore ideas is also an effective way of increasing contact between participants. Having several catering points to save on waiting in long queues can also help increase contact time.

Many people other than conference participants may be interested in the outcomes of a particular conference. Hence conference proceedings are a valuable tool for extending networks associated with the subject of the conference.

The Nature Conservation series of Surrey Beatty & Sons provides an example of the importance of a conference series in fostering the development of networks for conservation. The first of this series was based on a conference held in 1985 on the subject of the role of remnants of native vegetation in conservation (Saunders *et al.* 1987). It was initiated because CSIRO Wildlife and Ecology was developing a research programme on the conservation importance of remnant vegetation and needed to establish what research was being conducted at that time. The Western Australian Department of Conservation and Land Management was involved with the organization and the conference itself lead to the linking of a number of individuals and research establishments both nationally and internationally. The proceedings of the conference were published after a rigorous refereeing process in the same way that papers for international journals are refereed, with referees being able to reject papers which are not of suitable standard. This process depends on the development and maintenance of an editor's network, which is essential for the maintenance of scientific standards.

Since then there have been four other conferences which have all developed as an extension of the previous conference. The most recent with published proceedings was held in May 1994 in conjunction with the University of Auckland (New Zealand), World Wide Fund for Nature (Australia) and the Western Australian Department of Conservation and Land Management. It explored the themes that networks of people are our conservation force and networks of other organisms are our conservation resource (Saunders *et al.* 1995). This conference fostered the development of a number of networks for conservation as well as publicising networks in existence.

MEDIA

The media is a powerful force for providing information and for effecting change in society. If ecologists and conservation biologists are to be successful in spreading their important environmental messages they should try to work closely with the mass media. Ehrlich (1993), Jacklyn (1995), Kelly and Martinovich (both this volume) provide information of use to those needing support in dealing with the media.

The media can be very useful in developing and maintaining networks. Many scientific papers can be the basis of raising environmental awareness among the community and the many politicians who get much of their background information from the media (Comben, this volume). Once a paper has been published in the scientific literature it has been subjected to peer scrutiny and is available in the public domain. It is useful to prepare a press release highlighting the important findings and their environmental implications and release it to the major media outlets. This can lead to the development of networks involving journalists and broadcasters who need information to feed an industry with an insatiable appetite for such information.

POLITICIANS

Politicians make decisions on behalf of society on the way the environment, including natural resources are managed. Comben (this volume) explains the constraints under which politicians operate. Politicians are, therefore, important targets for environmental education and should be included directly in networks for education and action. It is not sufficient to rely on them picking up their environmental information from the media and certainly not from the scientific literature; they need to be dealt with directly. Contrary to popular belief, politicians are also human and do respond to invitations to see field work in progress or attend environmental briefings if they can find the time or the invitations are couched in an interesting way that attracts their attention against the many competing demands for their time.

CONCLUSIONS

Given the environmental problems facing us and the pivotal role that ecologists and conservation biologists have in research and application to address these problems, the development and maintenance of networks are essential. Networks are essential to ensure that environmental messages are disseminated as widely as possible to influence changes to more environmentally sustainable land uses and to ensure that research results are combined with other relevant research to affect the conservation of natural resources. Each ecologist and conservation biologist should endeavour to develop and maintain appropriate networks to see that their research is used to improve the management of our environment.

ACKNOWLEDGEMENTS

I am grateful to the following: Ray Wills and Richard Hobbs for the invitation to present this material at the Ecological Society of Australia's 1996 conference in Townsville; Richard Hobbs and Libby Mattiske for constructive criticism of this chapter; and to the many people I have been associated with who have helped shape my ideas on the importance of networks for conservation.

REFERENCES

Atkins, L. and Hobbs, R. J., 1995. Communicating research: How well are we really doing? Pp. 649–52 *in* Nature Conservation 4: the Role of Networks ed by D. A. Saunders, J. L. Craig and E. M. Mattiske. Surrey Beatty & Sons, Chipping Norton.

Commonwealth of Australia, 1995. Native vegetation clearance, habitat loss and biodiversity decline. Biodiversity Series, Paper No. 6. Department of the Environment, Sport and Territories, Canberra.

Commonwealth of Australia, 1996. Australia: State of the Environment 1996. CSIRO Publishing, Collingwood.

Doley, A., 1995. Networks needed by individual farmers: a personal perspective from the northern wheatbelt of Western Australia. Pp. 241–50 *in* Nature Conservation 4: the Role of Networks ed by D. A. Saunders, J. L. Craig and E. M. Mattiske. Surrey Beatty & Sons, Chipping Norton.

Ehrlich, P. R., 1993. Communication: how can ecologists get their message out? Pp. 295–301 *in* Nature Conservation 3: Reconstruction of Fragmented Ecosystems ed by D. A. Saunders, R. J. Hobbs and P. R. Ehrlich. Surrey Beatty & Sons, Chipping Norton.

Gilmour, J. and Saunders, D. A., 1995. Earthwatch: an international network in support of field research. Pp. 627–33 *in* Nature Conservation 4: the Role of Networks ed by D. A. Saunders, J. L. Craig and E. M. Mattiske. Surrey Beatty & Sons, Chipping Norton.

Hobbs, R. J. and Saunders D. A. (eds), 1993. Reintegrating Fragmented Landscapes: Towards Sustainable Production and Nature Conservation. Springer-Verlag, New York.

Jacklyn, P. M., 1995. Using the mass media in networks. Pp. 589–93 in Nature Conservation 4: the Role of Networks ed by D. A. Saunders, J. L. Craig and E. M. Mattiske. Surrey Beatty & Sons, Chipping Norton.

Lefroy, E. C., Hobbs, R. J. and Atkins, L. J., 1991. A Revegetation Guide to the Central Wheatbelt. Western Australian Department of Agriculture, Greening Western Australia and CSIRO, Perth, Western Australia.

Saunders, D. A., 1990. Problems of survival in an extensively cultivated landscape: the case of Carnaby's cockatoo *Calyptorhynchus funereus latirostris*. *Biol. Cons.* **54:** 111–24.

Saunders, D. A., 1996. Does our lack of vision threaten the viability of the reconstruction of disturbed ecosystems? *Pac. Con. Biol.* **2:** 321–26.

Saunders, D. A., Arnold, G. W., Burbidge, A. A. and Hopkins, A. J. M. (eds), 1987. Nature Conservation: the Role of Remnants of Native Vegetation. Surrey Beatty & Sons, Chipping Norton.

Saunders, D. A., Craig, J. L. and Mattiske, E. M. (eds), 1995. Nature Conservation 4: the Role of Networks. Surrey Beatty & Sons, Chipping Norton.

Saunders, D. A. and Ingram, J. A., 1995. Birds of Southwestern Australia. An Atlas of Changes in the Distribution and Abundance of the Wheatbelt Avifauna. Surrey Beatty & Sons, Chipping Norton.

CHAPTER EIGHT

Ecotourism in communicating ecology

PENNY VAN OOSTERZEE[1] and NOEL PREECE[1]

Tour operators possibly educate more people about Australia's ecology than all ecology lecturers combined. In addition to this, a case could be made that much of tourism in Australia, from international marketing through to taking a nature-based tour, is based on the country's ecology. The tourism industry and tourists themselves should be major promoters of ecological research and users of its results. Yet one of the most obvious deficiencies of tourism agents and operators is the level of education and communication skills. Misinformation on biodiversity, ecology and management is particularly common.

The main goal of communicating ecology is conservation of ecosystems. To communicate ecology effectively, passion and politics must be put back into the equation.

ECOTOURISM IN COMMUNICATING ECOLOGY

THE idea of running tours with an ecological bent was first suggested to me by a friend as we were travelling along the west MacDonnell Ranges in central Australia. As an ecologist myself, I remember treating the idea with considerable disdain. With the higher goal of doing a Ph.D. the thought of becoming a lesser mortal, leave alone a lowly tour operator was nothing short of preposterous. Nevertheless, the thought had stolen into my head, and I began to take notice of the information that tourists were given when they visited Alice Springs.

It was actually the view from Anzac Hill that was the catalyst to setting up Desert Discovery — before the term ecotour was even coined — now Discovery Ecotours. The view is inspirational, and tells a terrific geological story spanning two billion years. Backed-up by the fossils that you can find in the area, it provides a glimpse into the evolution of life and the Australian landscape that is simply mind-boggling.

Ten years ago, however, most people were shown little more than the roofs of the new shopping complex and the new set of traffic lights. In fact visitors left the evocative centre of Australia knowing little more than when they came. This was a tragedy. I have always believed that if people were inspired by what they were looking at — the red rocks, some older than life itself — and beyond the vast plains of golden spinifex, uniquely Australian with their teeming populations of wildlife ruled by termites and lizards, and tiny marsupials that would rip your thumb off — surely if people understood this, they would want to save it.

Stephen Jay Gould says:

"We cannot win this battle to save species and environments without forging an emotional bond between ourselves and nature as well — for we will not fight to save what we do not love" (Orr 1992).

I am of course assuming that the main goal of communicating ecology is conserving ecosystems even though I am aware that not all ecologists are necessarily conservationists.

Communicating ecology has a lot to do with communicating awe, mystery and love. Yet science often works inversely to passion. It is ironic really because, in the words of David Orr, "science, at its best, is driven by passion, and emotion. And we have emotions for the same reason we have arms and legs: they have proven to be useful over evolutionary time. The point in either case is not to cut off various appendages and qualities, but rather to learn to co-ordinate and discipline them to good use" (Orr 1992).

The main reason ecology and environment are not potent national political issues is surely because of the failure of scientists to communicate their passion adequately to the public. As ecologists we need to keep in mind that it was politics and economics that allowed science to flourish in the first place. Now more than ever before it will be politics and economics that allow it to merely exist. The decisions to conserve biodiversity will be political; the decisions to set aside representative areas will be political; the decisions on funding for ecologist's jobs will be political.

To be effective, communicating ecology should be a dialogue between science and politics. In the current climate of the 30 second

[1]EcO$_2$, G.P.O. Box 381, Darwin. Northern Territory, Australia 0801.

Pp. 33-35 *in* ECOLOGY FOR EVERYONE: COMMUNICATING ECOLOGY TO SCIENTISTS, THE PUBLIC AND THE POLITICIANS ed by R. T. Wills and R. J. Hobbs. Surrey Beatty & Sons, 1998.

grab, timidity and the professionally induced fear of lacking rigour will confine ecologists to quibbles of great insignificance. In today's political environment, this "more-rigorous than thou" attitude is in fact unreasonable, and, ironically, unscientific, because it does not analyse the means of achieving the goals of ecology and its interrelationships with the rest of the world.

What has this got to do with tourism? Tourism is driven by politics and economics. It is Australia's fastest growing industry, generating more than $13 billion in export earnings. The industry accounts for one in every eight export dollars earned by Australia and employs over half a million people. Forecasts show the industry's growth is set to continue. No other industry in Australia can boast such an impressive performance nor look forward to such an optimistic future. And if ever there were a major industry dependent on ecology and natural environments — at least in Australia — it is tourism.

Let us take a look at some economic values of World Heritage Areas since they were nominated at least partly as a result of the findings of ecological research. (I might mention here that I do not hold with the false distinction being made between tourism and ecotourism. I define ecotourism as all visits that focus on nature appreciation and the associated infrastructure, park facilities, tours, accommodation, airlines that bring tourists to the area etc., that supports these visits — that is, tourism.) Recently there have been some studies done of their economic value from tourism to regional economies. For the Great Barrier Reef, the number of tourists in 1991/92 was estimated at 2.2 million with direct expenditure, including travel, of $1,080 million dollars. Commercial fishing in comparison is worth a little over a tenth this value at $128 million. Management costs accounted for $18 million. Annual expenditure on research by the Great Barrier Reef Marine Park Authority and the Australian Institute of Marine Research was $19.4 million.

Approximately 1.8 million tourists visit the Wet Tropics World Heritage Area each year, with a direct expenditure of $377 (incidentally, logging, which ceased in 1987, generated a gross revenue of only $25 million per annum (Driml 1991)). Funding for management amounted to $12.05 million in 1991/92. In contrast, broad figures for research for last year, amount to around $2.0 million ($805,000 for research, inventory and monitoring and $1.572 million for research and management on threats to the World Heritage Area, in press for the WTWHA1995/96 annual report).

The number of tourists visiting Kakadu National Park in 1990 was around 240 000 with direct expenditure of $122 million. Funding for management is about $11 million and last year money spent on research and mapping amounted to $138,000 not including the 34 research permits granted.

This is "in your face" evidence of the economic benefits of natural areas. It also reveals the inequality of the funding that research receives if you consider that much of the value of those places is based on that research. I am not suggesting that tour operators should be paying for ecological research. I am simply stating the obvious — that, as ecologists, we should be using these economic realities to convince politicians of the value of ecology. As ecologists we should be harnessing tourism as a force for ecological research and conservation, and seeing it as an avenue to distribute the results.

Daily, there are hundreds of tour operators in effect teaching thousands of people about Australia's ecology. There are around 50 nature-based operators in the Wet Tropics alone. Because a tour occupies all the senses, it can get important ecological messages through to more people than Australia's 50 tertiary institutions.

Here is just a couple of responses we have had after running tours that link up with ecologists:

"I don't say enjoy because the world is not comprehensive enough; it goes beyond that. Thought provoking! I mean out there evolution became reality and survival the common denominator." *Judith Irani.*

Another: "After three days in the desert, I learned more about the natural history, geology, ecology and original inhabitants of Australia than I had absorbed through many years of formal education." *Maria Prendergast.*

People are actually hungry for information from scientists, but they cannot find them. Finding what research is going on at Kakadu, for instance, is one of the most difficult things to do. In addition, the ecological research being carried out in Kakadu is the only thing that is not really being interpreted to the general public. The park staff do, however, run a terrific course for tour operators and just about every tour operator avails themselves of it, even though it is not compulsory. Many other organisations could take a leaf from their book. But even in this course a discussion of the ecological research is done as a rather dry lecture by the coordinator of the research and not by the ecologists themselves.

This is important. For example, one of the persistent problems, with respect to misinformation in the Top End, is the interpretation of fire. I suspect that if tour operators were taken out, together with traditional owners, where the research is occurring, and shown by the ecologists doing the research the different habitats and the results of various burning regimes they would begin to truly understand the issue.

How do we communicate? There is plenty of advice about this in this volume. All I would like to emphasize is that ecologists need to learn to speak clearly and boldly. Ironically, we should not let the facts get in the way of a good story. Please understand me, we should be accurate but not boring. And from my own hard-won experience, a few pointers: we need to give a little of ourselves; we need to watch the arrogance of the scientist; and we need to have at least a passing understanding of other disciplines.

We do not have the luxury of saying "I do not have time".

The results of diminishing the distance between scientists and people can be astonishing. It is amazing the response you get when you put passion back into science. And scientists are actually good at it. A bunch of examples include Tim Flannery, Ken Aplin, Steve Morton — you can all think of more. The excitement of watching John Long see a perfect lungfish fossil, of watching Michael Archer tenderly stroke the vertebrae of a newly found giant emu, of Colin Limpus discovering an internationally significant population of turtles, is palpable.

Perhaps it is more than a coincidence that most of these people have relatively healthy funding.

REFERENCES

Driml, S. M., 1991. Research needs for rainforest economics. *In* Tropical rainforest research in Australia ed by N. Goudberg, M. Bonell and D. Benzaken. Institute of Tropical Rainforest Studies, Townsville.

Driml, S. M., 1994. Protection for Profit. Great Barrier Reef Marine Park Authority, Townsville.

Orr, D. W., 1992. For the Love of Life. *Cons. Biol.* **6:** 486–87.

Preece, N., van Oosterzee, P. and James, D., 1995. Two Way Track: Biodiversity Conservation and Ecotourism. Biodiversity Series, Paper No 5. Biodiversity Unit. Department of Environment Sport and Territories.

CHAPTER NINE

The politics of the media and ecological ethics

J. B. KIRKPATRICK[1]

Ecologists are scientists who attempt to test various possible stories about a reality that they believe ultimately exists. They feel that they have invalidated the perpetual economic growth myth. Journalists are communicators, driven by the need for audiences and the "growthist" world-views of their employers. The construction of "news" is such that issues are polarized and facts might as well not exist. The differences in training and culture between ecologists and journalists makes accurate reporting of ecological work difficult, especially in the frequent context of differences of opinion between ecologists. Ethical behaviour on the part of ecologists in this confusing situation is suggested to involve: honesty; concentrating on the ball, not the player; negotiation between proponents to isolate areas of disagreement and means of resolving such disagreement; playing the debate in open, rather than applying surreptitious pressure.

INTRODUCTION

ECOLOGISTS and journalists are very much engaged in politics, which is the art of resolving conflicts without killing or maiming anyone. Ecologists and journalists both have important roles in the global society. It could be said that ecologists discover news and journalists propagate it. The essence of the ethical position that I like to assume for these two professions is that ecologists discover problems and opportunities that, when solved or taken up, lead to an increase in happiness for people and other living things, while journalists communicate the problems and opportunities to those who can do something about them.

This chapter presents my perceptions of the manner in which ecologists and journalists operate in contemporary Australian society, and discusses some aspects of ethical behaviour in situations in which ecologists engage with the media.

The chapter is based on my quarter of a century experience of interaction with the media on ecology-based issues, and my personal observation of media outcomes, rather than on careful collection and analysis of data. Despite the methodological deficiencies of my approach, I hope that the chapter might aid those ecologists whose consciences impel them to become public, and, perhaps, provide a stimulus for more rigorous research.

MYTHS

All societies have widely-accepted value systems that implicitly or explicitly regulate their activities. These value systems are not necessarily consistent with the long-term survival of a society, as witnessed by the many civilizations that have collapsed because the interactions with the environment that flowed from their values proved to be unsustainable. I label such maladaptive value systems myths.

Among the many myths that underlie the activities of our society, the most influential is that which I call "growthist". This myth is that it is both possible and desirable to increase material production indefinitely on a finite planet (e.g., World Commission on Environment and Development 1987). Our institutions and activities have evolved to perpetuate and enable the expression of this myth. They are appropriately, in this context, focused on short term outcomes and the facilitation of change.

A second, possibly less pervasive, and certainly more recent, myth is the post-modernist intellectual construct. This myth sees only perceptions and interpretations, that relate to a text, or a planet, which has no single meaning. One of its effects is to increase the alienation of people from nature.

ECOLOGICAL REALITY

Ecologists see their work as fitting within the paradigm of science. Their particular interest is the interaction between organisms and their environment. Science is concerned with the small subset of human perceptions that are susceptible to invalidation through observation or experiment. Although many scientists become emotionally attached to one of the many stories, or theories, that can explain sets of observations about reality, their mission is to poke around in their weak spots, searching for invalidatory data, not to search for confirmatory data. Given that a large proportion of

[1]Department of Geography and Environmental Studies, University of Tasmania, Box 252C, G.P.O., Hobart, Tasmania, Australia 7001.

Pp. 36–41 *in* ECOLOGY FOR EVERYONE: COMMUNICATING ECOLOGY TO SCIENTISTS, THE PUBLIC AND THE POLITICIANS ed by R. T. Wills and R. J. Hobbs. Surrey Beatty & Sons, 1998.

important scientific stories arise from a synthesis of purposive and accidental observations, the absence of an hypothesis, deductively derived from a theory, is not a necessity for scientific endeavour. Whether the approach to story-development is through deduction or induction, scientists need creativity and the ability to logically process information. Their goal is not to determine the "scientific truth", but rather to develop better and better stories about a reality that may be essentially unknowable. Ecologists therefore suspect that there is a reality. Otherwise, why try to approximate it? They recognize that some stories about it are wrong, that others are possible, and that one or some of these others provide the best fit to available information. They are definitely not post-modernist in their outlook.

One story about reality that ecologists are confident that they have invalidated is that it is possible to increase the rate of material resource use indefinitely in a finite ecosystem. The biosphere is one such ecosystem. Thus, ecologists see the term "sustainable development" to be an oxymoron. Our only problem is that the prediction of behaviour in complex, partly chaotic, systems approaches the impossible. So, we have little or no idea of the timing of depletion or of the timing of ecosystem transformation to lower productivity levels.

The sociological structure of science is centred on publication in peer-refereed journals. The prestige of individual scientists relates, in descending order, to the perceived impact of their work on other scientists (often measured, inexactly, through citation indices), the volume of their work and the volume of the funds that support their work. Most of the ecological scientific work is highly reductionist in approach, testing ideas about individual interactions, rather than exploring whole systems. It is not value-free. Peers set the limits of behaviour and expression, which constantly shift in a process of social consensus. The ecological scientist may only be able to gain money to support work in an area that is defined as important by the granting body or peers. The individual ecologist who has more freedom might select an area for investigation because of deeply held values, that, like most values, are not susceptible to invalidation through the processes of science.

THE JOURNALISTIC WORLD

Although there are a few journalist who possess an high degree of understanding of science and its sub-branch, ecology, most journalists are not familiar with scientific paradigms and processes. Very few journalists are familiar with ecology as a scientific endeavour, although more have a vague idea of its importance through contact with the ideas of the conservation movement and "new age" proponents.

Most journalists rely for their livelihood on contractual arrangements with, or employment in, the mass media. In Australia, and most other parts of the world, most of the mass media is profit-oriented. The members of the boards of private mass media companies have value frameworks that support the growthist ideology. A capitalist economy is one of the major expressions of growthism (it also thrives in state socialism). This means that board members and executives are unlikely to look positively on any challenges to growthism, and are naturally more sympathetic to the development side of any development/ conservation debate. Employees and contractors who challenge growthism through their work in communicating news, information and ideas to the public are unlikely to have long or lucrative careers. Given that almost all politicians also support and promote this ideology, journalists critical of growthism who work in public media feel similarly constrained.

While growthism is seldom challenged directly in the mass media, conservation/development debates are frequently reported and analysed. There are several possible explanations for this phenomenon, among which the economic explanation seems strongest. The profitability of the private media depends ultimately on its audience, which in turn attracts its advertisers, who in turn provide the profit. If a large proportion of the population are concerned about a particular issue, and if the issue lends itself to sufficient dramatization to attract the unconcerned, large audiences can be created.

Perceived potential profit from audience gain must often outweigh the perceived potential losses which would be suffered by the members of mass media boards and the mass media companies as a result of media exposure of development/conservation issues. The magnificent river theatre of the Franklin blockade gained outstanding national media coverage, as did the whole of the on-going debate between conservationists and the Hydro-Electric Commission. It was only in Tasmania that the mass media owners had any great economic interest in this particular development. Debates over native forests, which provide much of the raw material for newspapers, have not been as prominent in the media. Part of the explanation of this difference could relate to the relative potential impacts of conservation on the economic interests of the media companies and their boards. According to Burton (1996) the Australian forest industries have attempted to keep the conservation/development debate low key, as a result of market research that indicated that any news in this area activated public concern.

The power, prestige and income of a journalist seems primarily related to the size of their audience, and only secondarily related to the quality of their work. If quality is related to accuracy and depth of analysis in reporting, it is seldom possible for journalists to be able to achieve highly on both criteria. The most accurate and insightful reporting tends to occur in the audience-challenged media that is owned by the public, and in a few high quality, but low circulation, private publications.

Ecologists are reported in the media in two distinct situations. The public has a sufficient interest in nature to be attracted to interesting information about its functioning, especially in television and print. A strong story-line and good visuals are necessary to get such exposure. These are often provided by ecological research. In the second situation ecologists are involved in the "news", usually as proponents in a debate involving a potential development.

The nature of "news" varies somewhat between types of media. However, there are several elements that appear both universal and temporally stable. Disasters and tragedies, preferably involving substantial loss of human life, appear to be compulsory. The viewing of these may make people feel more secure or lucky in a world in which the only personal certainty is death. Political conflicts seem to be as compulsory an element as tragedies, the sport results and the weather. Reporting of crimes and other misbehaviours, and their consequences, is a close to universal component in the print media, but these have to be spectacular to penetrate the electronic news. Good news stories are optional in the electronic media. They are more frequent in the print media. They tend to the lachrymose.

Because "news" is trying to attract an audience, political debates are dramatized. The reality of most development/conservation debates is that there are more than two sets of views, and complex patterns of agreement and disagreement. The agreements are not usually dramatic enough to be reported. That is why few people know that most bills that pass through our various parliaments are passed unanimously. If there are more than two positions in relation to an issue, the best "news" usually results from dramatization of the extremes.

Most journalists work under severe deadlines, with little time for research. It is not surprising, therefore, that such research that they normally do consists of a search for extreme points of view, and that it is these points of view that dominate reporting. One is almost forced to be a post-modernist to be a reporter.

One major consequence of reporting the views and "facts" of the extreme protagonists in debates is that the public tends to find it difficult to identify what is true and what is not. A second consequence is that "facts" tend to be flexible. For example, the media reported an exponentially increasing number of potential job losses from National Heritage listing of forests in Tasmania at the time of a woodchip Environmental Impact Statement (EIS), even though the later figures exceeded the number of jobs in the forestry sector of the economy (Fig. 1; Kirkpatrick 1987). The reporters may not have noticed the change in magnitude. It certainly was not brought to public attention.

The more recent media conflict over the interaction of the silvicultural techniques used by the Western Australian Department of Conservation and Land Management (CALM) with the phascagole further illustrates "fact" shift. A university Ph.D student published a paper in a refereed journal, which was reported in the *West Australian* on April 1, 1996 in the following manner: ". . . Susan Rhind, warned that CALM plans to allow logging in corridors between logged areas within 20 years could result in local extinctions of hollow-dependent species such as the phascagole . . .". In the same newspaper, on April 6, 1996, an article reported reaction to a letter criticizing Rhind's work that was sent from CALM to the chief executive officer of Murdoch University. In this article it was stated that her: ". . . research, made public last month, claimed that logging practices would cause phascagoles to become extinct in the South-West." Local extinction became regional extinction in only five days.

Apart from the consequences of the dichotomization process, the accuracy of reporting often leaves a little bit to be desired, as evidenced by the "we were wrong" sections in newspapers. One prominent Tasmanian newspaper usually fails to accurately report the census results, although its sports results are generally reliable. Most reports in the media on matters on which I have had inside knowledge, contained factual mistakes, important omissions or inappropriate emphases.

ECOLOGISTS IN MEDIA-MEDIATED DEBATES

Given the rapid changes in land use and landscapes in our country (State of the

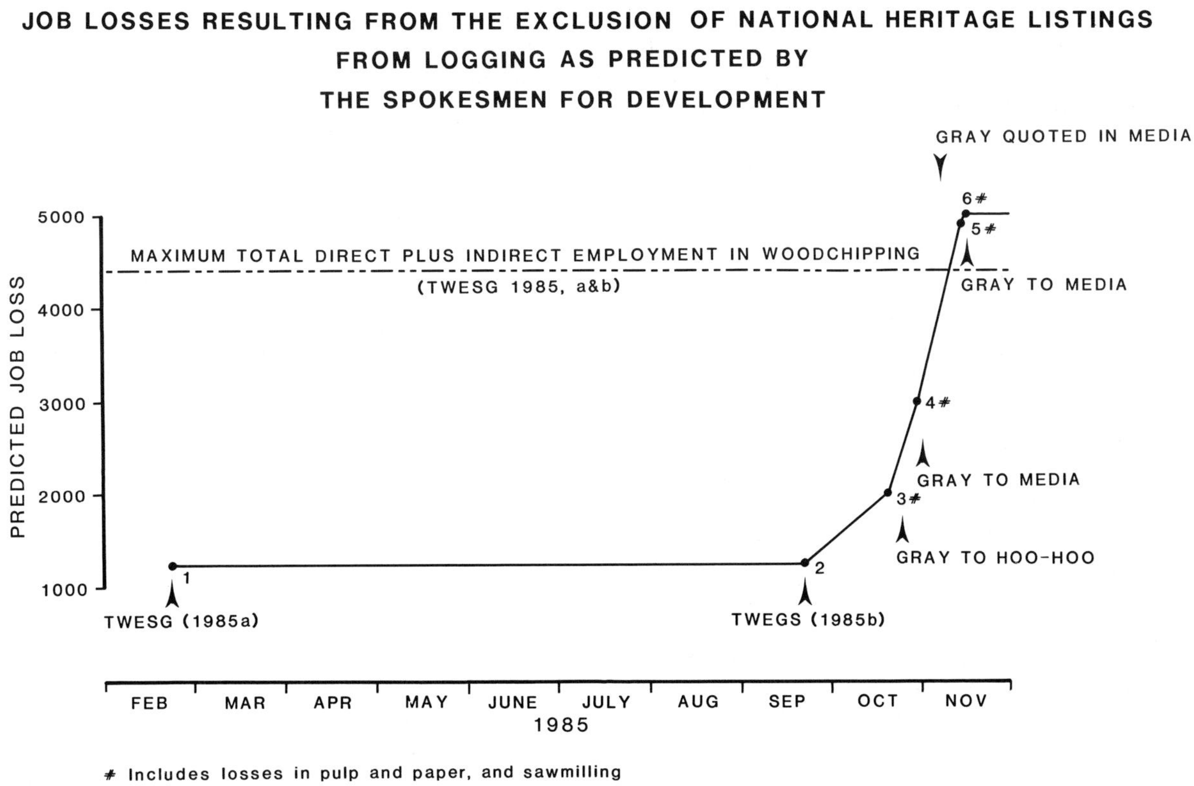

Fig. 1. Job losses predicted by development spokespeople during the final stages of decision making on a woodchip export licence for Tasmania (Kirkpatrick 1987). The horizontal line indicates the number of job losses predicted in the EIS, which was prepared by the developers.

Environment Advisory Council 1996), it is not surprising that ecologists often collect data that indicate that an actual or proposed development activity is having, or could have, deleterious effects on organisms or ecosystems. If ecology is not to be equated to measuring the dimensions of the deck chairs on the Titanic, the public must be made aware of such data and its implications. This necessarily means some exposure in the mass media.

Ecological issues are frequently the subject of conflict between ecologists and non-ecologists, for example, between a developer who believes that the economic return from their canal estate is much more important to society than its potential contribution to the endangerment of a species of marine mammal. This example is simply a values clash, resolvable through the political process. There are, however, also conflicts between ecologists. In many conservation/development debates ecological researchers have been protagonists on both sides. Recently, there have been at least two examples of such debates between ecologists that have gained a high profile in the media, one to do with the potential impact on platypus populations of the draining of the impoundment which flooded Lake Pedder, and one to do with the potential impact of forest management practices on phascagoles in jarrah forest.

In the first case, a University of Tasmania academic suggested in the mass media that the draining of the impoundment could lead to a reduction in platypus populations, while another, in the same institution, heavily criticized this opinion, also in the mass media. At least one letter, from a politician, was written to the institution suggesting that action against the platypus depletion hypothesis proponent should be considered.

In the phascagole case the *West Australian* (April 1, 1996) reported that: "CALM science and information director Jim Armstrong said none of Ms Rhind's claims was valid or substantiated in the paper by unequivocal data." Jim Armstrong was also reported as saying that: "The paper was flawed because it did not have a hypothesis, data presented in the standard format . . .", and ". . . it was regrettable that personal opinion was allowed to override objectively gathered data". In the same newspaper, on April 6, 1996, an article reported reaction to a letter criticizing Rhind's work that was sent by Dr Armstrong to the chief executive officer of Murdoch University.

In this article Dr Armstrong was reported as writing: "... Ms Rhind has an earnest desire to prove that logging in the jarrah forest is detrimental to the brush-tailed phascagole, and that this opinion has been allowed to overshadow the need for objectively gathered data, the hallmark of soundly-based science." The ultimate outcome of the phascagole case was an apology from Susan Rhind to CALM "... after the department (CALM) had provided photographic evidence that identified the error." (*West Australian* Saturday, May 4, 1996). There was no attempt by the *West Australian* to identify the parts of her research that retained validity.

Many ecologists have experienced distortion of their results and professional opinions in the mass media, and/or pressure on our organizations from people or organizations that do not like our results or their interpretation, and/or conflict with our ecological colleagues on public issues, as in the cases above. Distortions and pressure tend to be associated with issues that carry a high emotional load, such as Lake Pedder and the future of the Western Australian forests.

There are a few behaviours that can aid ecologists in interactions with the media in relation to such conservation/development debates:

1. media exposure will be maximized by maximum dichotomization of the issue — this is both an opportunity and a trap;
2. it is very important to correct mis-reporting, even of minor details — "we were wrong" columns are very small but very powerful — they give a high degree of veracity to your case;
3. do not say anything to the media that you are not confident is strongly arguable, by yourself;
4. qualify your statements to indicate degree of certainty — words and phrases like "probably", "likely to", "a strong likelihood that" are extremely powerful — you use them liberally in your papers — do the same in media interaction;
5. do not attack your opponents on a personal level — it may be extremely tempting to say things like "developers have thick necks because they spend their leisure hours biting the heads off baby wombats", but resist the temptation;
6. allow your emotions to be apparent, especially in the electronic media — emotions reinforce a logical case;
7. use your professional society if you are being badly treated in the media, and help it to develop policy — comments have much more strength from a society than from an individual.

ETHICS IN MEDIA DEBATES

While it is proper that members of the Society who have different professional opinions on an issue should engage in public debate, there seems to be a need for a code of conduct for such engagement. In my opinion, the first element in such a code of conduct should be that any opinions given should be honestly held and their limitations adumbrated. Any set of information can be interpreted honestly in a variety of ways, depending on the theoretical framework of the interpreter, their wider experience and the context in which they place the information. Statements of professional opinion should be presented as such, not as "scientific truth", which does not exist. The basis for these opinions should be clearly stated — is it the gut feeling of the communicator? Is it the communicators interpretation of a limited set of data? Is it a deduction from theory?

I believe that a second component should be the clear separation of personal opinion on an issue from opinions on the character and motives of ones opponents. Personal vituperation, when mixed with professional opinion, can only bring ecologists into disrepute.

Communication between the proponents of different views seems to me to be a desirable third element in a code of conduct. Proponents should be able to jointly identify their areas of scientific disagreement and jointly suggest scientific means for resolving them. If they cannot do both of these things, one can only assume that at least one proponent is disagreeing with the other on a non-scientific basis. Ecologists would gain respect from the public if they were able to say: "We disagree on the meaning of these data. However, we have agreed that this question could be resolved by the million dollar research programme detailed below" or "consideration of the information available to both of us, led us to agree on the following: ...".

A fourth element in a code of conduct should, I believe, require that debate between ecologists on issues should be open and between the proponents, conducted by means of one or all of the scientific literature, verbal debate and the media.

The first two elements in the code of conduct suggested above also obviously pertain to debates that involve ecologists on only one side.

The fourth element in the proposed code of conduct would not prevent non-ecologists attempting to gain advantage in a debate by political means. Universities usually prove capable of maintaining academic freedom. I wonder if government departments and other employers of ecologists would always demonstrate the same resolve. I believe that professional societies should have a policy of publicly defending members if such a resolve is lacking and the member has behaved ethically.

The ethics of behaviour in public disputes may seem relatively clear. Ethical behaviour becomes much more difficult to define before an issue has become public. It has been suggested by other authors in this volume that ecologists have a responsibility to make the public aware of environmental problems that have been revealed by their work, and a further responsibility to ensure that the problems are solved. This raises the possibility of sins of omission. While ecologists can be very good at identifying and understanding environmental problems, their skills and training do not lie in public relations and politics, and most are not emotionally suited to these activities. Moreover, if research ecologists try to become communicators, their vital research effort will necessarily be reduced.

Less than one thousand Australians cannot be held totally responsible for our environmental future. Nevertheless, it seems ethical for ecologists to communicate the problems and solutions they identify to those better adapted to the worlds of communication and decision-making, rather than confining their communication of insights to their colleagues and students. The media and/or conservation organizations can often take the results of ecological research and translate them into wider communication and action. It is their responsibility to do so. It might even be suggested that an increase in ecological literacy in these two areas, to the point at which a substantial number of journalists and conservationists became capable of understanding papers in ecological journals, would reflect ecologically ethical behaviour on their behalf.

REFERENCES

Burton, B., 1996. Sabotaging environmentalism with dirty tricks. *The Daily Planet*, **No. 43:** 8–9.

Kirkpatrick, J. B., 1987. The growthist/greenie ritual conflict and the nature of the argument — the example of the woodchip export EIS. Pp. 312–18 *in* Readings in Australian Geography ed by A. Conacher. IAG (WA) and Department of Geography, University of WA, Perth.

State of the Environment Advisory Council, 1996. State of the Environment Australia 1996 — Executive Summary. CSIRO, Collingwood.

World Commission on Environment and Development, 1987. Our Common Future. Oxford University Press, New York.

CHAPTER TEN

Ecology TV

KARINA KELLY[1]

Ecology dropped down the political and media agenda in the 1990s as did many science issues. How can ecologists sell their story and still sleep at night? This chapter takes a look at the concepts of scientific objectivity, the public's rapidly changing attitude to science, the status of the expert and how the media works in choosing stories. It argues that the fault is not all the media's when it comes to the coverage a story gets.

ECOLOGY AND THE MEDIA

SOMEONE once said to me early on in my media career that if there is a choice between a conspiracy and a stuff-up — you should choose the stuff-up every time. Of course my advisor was not being original, but the sentiment is a good one when you are thinking about the media. In a country of increasingly centralized media ownership and an ABC under serious threat of being slowly strangled to death because of lack of funding, you might be forgiven for thinking that your research project or burning cause has been ignored or badly covered because of a conspiracy. But a stuff-up is much more likely and consider this; the stuff-up might be yours.

In this chapter, I will go through how science has changed in the public's mind — and speculate about why that might be so. I will touch briefly on what makes a good story and finally what you are competing against when you try to get your story covered.

For international readers and those Australians who have not watched television for the last twelve years, I should explain that Quantum is a half hour science magazine programme on ABC television which now airs on Thursday nights at 8.00 p.m. For all of its twelve years, Quantum has been one of the most popular Australian-made programmes on ABC.

Some people in the broadcasting field find this surprising — indeed early on in his stewardship former ABC boss, David Hill saw me at a programme launch and said "Ah, yes, Karina . . . Quantum . . . Rates better than it deserves to." To David's credit, though — while he may have preferred to watch the football — he was not about to argue with success — our popularity and the esteem with which we are held in the science community — was enough to ensure our continuity.

These days we live in more uncertain times — we at the science unit have been unable to get any strong show of support for specialist science programme makers — indeed the future of the whole ABC hangs in the balance. Most of the people in this room would be paid from the public purse, and it is not very fashionable is it? It is as if what we are doing is not quite proper — because we are not out there making money for Australia.

All of this does beg the question what are we making money for? At the risk of appearing childishly simplistic, why did we come together into communities in the first place? Surely part of the reason is because together we could achieve more than we could individually. Of course we hope for greater prosperity and security, but we also want a sense of belonging — perhaps a sense of achievement that our community could look after the less fortunate.

Conserving the environment is part of that understanding that "there's something out there more important than money". Unfortunately in my opinion, science has been sold too much lately as something that could make us money. Indeed a great deal of our move in the last century away from religion and old world ethics has gone hand in hand with the rise of science. In the new scientific world (communist or capitalist) we were going to be more rational — less superstitious . . . and with science to hold our hand we were going to become more prosperous — our children would do better than we have and the promised land seemed almost in sight.

[1]Quantum, ABC Television, P.O. Box 9994 G.P.O., Sydney, New South Wales, Australia 2001.

Pp. 42–45 *in* ECOLOGY FOR EVERYONE: COMMUNICATING ECOLOGY TO SCIENTISTS, THE PUBLIC AND THE POLITICIANS ed by R. T. Wills and R. J. Hobbs. Surrey Beatty & Sons, 1998.

When I was finishing school and studying at university in the 1970s we were all told that the paperless office and the age of leisure were just around the corner. People would work ridiculously short hours because robots were going to look after us. Well, I suppose for all the unemployed people of Australia — it is the age of leisure but not one they welcome or celebrate. No, the promises of the seventies turned out to be empty. I was sitting in a taxi in Melbourne on my way back to the airport for the final leg of what would be a fourteen hour day. The taxi pulled up next to a dusty old monument erected in the early part of this century and chiseled into the stone were the words — "to commemorate the achievement of the eight hour day." It felt like a very long time ago.

So where is all this leading? I think to many members of our community it has led to a feeling of disenchantment. People can no longer fix their own cars, because there is a computer inside. They do not grow their own food or make their own clothes. These days an increasing number do not even make their evening meal. And while this keeps the economic cogs turning, and offers the producers economies of scale, it has the disadvantage of making people feel alienated — less in control of their daily lives. It makes them feel like a very small cog in a very large machine. To some of them, science has been the engine of this change in their perceived status or self esteem. Without science and engineering, where would the new products come from that must replace the old so that companies can make us go out and buy again.

In general then, most areas of science have been vital to this seemingly inexorable change we witness — which makes the advice of the parents increasingly irrelevant to their children. We live with uncertainty, we live with booms and busts, we believe like the addicted gambler that against all the odds, we will win the prize of economic expansion and someone else will lose.

There is no doubt that science is the most successful human endeavour of the last two thousand years but we are in danger of losing the strength it brings unless we can counter a growing backlash against science. And here I would like you to consider something heretical; that some of that disillusionment is justified.

Right now science is under enormous pressure to show that it can bring home the bacon — that it can pay dividends. What I have noticed over the last eleven years is that as science has become more commercialized — the CSIRO and Universities told to go and raise money from industry — the public's trust in science has dropped. Now you can interpret that as the public being led astray by astrologers and crystal peddlers or you can interpret it as a common sense reading that companies are there to make money and fully funded government scientific institutions are not. Perhaps then, while they might not always be efficient, the government institutions have no reason not to serve the public good.

In the meantime, the anti-science movement is getting some scientists riled. No-one gives up power or status willingly. In fact, I am already sensing a "backs to the wall" attitude among some of my scientific contacts which could lead to disaster. The one thing scientists must not do is hide behind their status as experts . . . or grumpily reject those people who question the worth of "western science" as you might have heard it called. You will also have heard the argument that it is one of a number of "value systems" that are equally valid.

It is important for the supporter of science to argue the case with good humour and recognize weaknesses in the system honestly. We know that science is not a religion, nor does it give philosophical guidance. It is a tool — a marvelous tool — that allows us to understand the world but it is up to us to decide what we do with that understanding.

As a community we are a bit rusty on the ethics front — these days we prefer to believe that if we trust in market forces — if we can only roll that level playing field flat enough — we will not need to engage in messy and irrational discussions about what is right and wrong.

But there is a group of disciplines which offers great hope in redeeming itself in the public's eyes. Environmental science, atmospheric science, renewable energy research and ecology are all areas that offer this hope. And the public is behind you. I was recently asked to be a judge of the Reader's Digest Environmental awards which gave $25,000 to a community project trying to conserve and improve their local environment. Reader's Digest did a survey in which they found that the environment was as important to people as the economy. This is not the first survey to show this — and it will not be the last. Even the politically dry Howard Government is recognizing this with their grand environmental policy albeit dependent of their sale of Telstra.

So, if Reader's Digest is on track and the politicians are learning there is another track why your project has not got the coverage it deserves? It must be a media conspiracy, right? Well, no. I think you will find that most journalists are on side but they do need scientists who are prepared to stand up and be counted.

Some of you may have seen the coverage John Wamsley has received over recent years. He is famous for wearing a feral cat hat — and that certainly is a great attention-grabber. But James Woodford's report in the *Sydney Morning Herald* on the 13th April also showed other aspects of Wamsley's story which made him a story-teller's dream. He left school young at 16. Became a trainee metallurgist. Worked extra shifts and double time to buy rundown houses and do them up. By 24 he owned 13 properties. He went to university at 25 and emerged at 30 with a Ph.D in mathematics. He bought his first sanctuary in 1969 and fenced it in 1982. He has had great success in breeding endangered species — even if they are in what the wildlife people call very large zoos. If extinction is the problem then successful breeding must be part of the solution. His aim is to become the biggest private landowner in Australia. And while everyone else has been talking about extinction, he has been doing something about it. He might be abrasive but he is prepared to stick his neck out . . . and that gets him media coverage.

Now I am not advocating that we start a new fashion in feral animal hats, but I think you can see that his story is a good read. Unlike much of current affairs coverage, at Quantum we are not looking for conflict to sell a story (I like to think that what you take away at the end of a Quantum story is greater understanding). But we do need an angle. Why is this story different to any other? Why should I care about it? You care about it because you have been working on it for ten years — but why should I care? And if that sounds callous think of this. Every year my family contributes to Amnesty International, the Australian Consumers Association and The Australian Museum Society. We send money to the Smith Family, Community Aid Abroad, the Salvation Army. We subscribe to Science, The New Scientist and Scientific American as well as two daily newspapers. We all have a lot of things we care about. The problem today is not access to information but information overload. So if you want me to care about the extinction of the lesser spotted bandicoot — you had better show me that you are prepared to stand up and be counted that you are doing more than just what you are paid for.

The science must be just as rigorous. The peer-reviewed publications are just as vital. But it really is not true that you (the scientists) are wholly objective. We all bring our beliefs and philosophies to work with us. Good journalists worry every day about representing their subjects with integrity. Good scientists must make sure that their opinions do not affect their methodology. So long as you do not mistake opinion for fact, it is OK for you to air your opinions.

And if you will allow me to air one of mine, you need a good copywriter for some of the most powerful issues in ecology. For example, how often have you heard people use the phrase "economically sustainable development" instead of "ecologically sustainable development" and both of them are a mouthful. And the precautionary principle (that if you do not know, do not do it) which I think is a wonderful concept needs a new name — people look at me blankly when I use it — but the idea behind it is a radical departure from the conventional game being played which says anything is OK unless you can prove that it does harm.

Most of the people in our society get most of their information from television. Many use no other source of information. And here is a problem that we in television must try very hard to fight. It is not a conspiracy but rather the industry's belief of what makes "good television". Firstly television tells personal stories very well. You feel you know someone after you have seen a story on them. You might see them in their house, you might see them interact with their colleagues and family, you hear their opinions while watching their faces — and we all pride ourselves on being a good judge of character. The current affairs programme Sixty Minutes, is based on the idea of exploring an issue through the eyes of an individual or individuals. Television is so powerful when it is used this way that many people working in television do not want to see it do anything else. Television can do analysis, it can explore ideas, new ways of seeing the world but there is no question it takes more time and money and it takes more thought. It urgently needs to do these things because for many people, the box is their only window on the world.

Television is also driven by fashion and the fashion today is to deconstruct. I have even heard it argued at the ABC that for youth programming at least, the programmes should be constructed for people who "surf" television. They click on with their remote controls, and need to be able to understand immediately what is going on. To me, this sounds like we would be making programmes for people who are not watching us. To be deconstructed also means — to me — the end of reasoned argument. If every programme is deconstructed it becomes impressionistic — clips, visions, views. There is no room for a series of points being made that lead to the programme's conclusion.

But enough of my problems. What about yours? Public funding for everything is in short supply and will stay that way for several years. The great asset that ecologists have is that the public genuinely wants to see the environment preserved. So, in the short term, the future for ecologists lies in local, community-based support.

Deeper than that, it seems we have come full circle. First, the grand Eighteenth Century view that the world and all the creatures in it were created by God for us to use — a garden for us to play in. Then, beginning with Darwin and continuing through the Twentieth Century a growing unease about our role and importance. Evolution told us that rather than being the pinnacle of God's creation, we were just an accident, all the reasons for our superiority — language, tool use, culture, which we told ourselves made us different from the animals fell away. But as the Twenty First Century begins — I see a more confident self image for us. Our role is clear. We are here to tend the planet — to maintain species diversity and ensure that our future is sustainable. It must be conscious on our part. It is not going to happen by accident and in this broad sense we can again assume the role of the gardener.

I think the greatest challenge for us is that like a stuntman hooked on adrenaline, our society is hooked on change. Our whole economic credo is based on change — because it is based on growth. It is going to take a lot of talking and a lot of lobbying and a lot of very good science to convince our community that we should stop and think about how we change in the future.

CHAPTER ELEVEN

Communicating science effectively

MONICA MARTINOVICH[1]

INTRODUCTION

SCIENTIFIC research has changed radically over the past few years. Scientists have far less freedom to carry out their research. A great deal of time is taken up in trying to obtain funds to keep a project going and even more time is spent on accountability for the work being produced. This entails review after review.

How then, by spending even more time away from research communicating your research to politicians, the general public and "the media" does the scientist benefit?

By using good communication skills, and working within an effective communication programme, scientists gain credibility with the general public for their research programme. The general public actually understand the research work being carried out.

This also means that both business and government organizations are also aware of the scientists' work, which helps immensely when applying for funding, grants or sponsorship

Government departments and especially businesses are far more interested in good public relations, (good publicity) and/or monetary benefits, than the philanthropic view point of research. Unless of course that research directly helps them in some way.

In this chapter, I present my perspective as a professional communicator on the need for increased communication by ecologists, and provide some hints on how to do it more effectively.

REASONS FOR COMMUNICATING

To quote Cory Dean, a senior journalist from *The New York Times,* three main reasons for science communication are:

1. Public accountability, because most research is funded by tax dollars.
2. Providing scientific information with which the public can make important decisions about matters which impact on the life of the community.
3. The need for a scientifically literate workforce.

A scientist who skillfully communicates gains maximum benefit from any interaction he or she may have with the media. Journalists will always use a scientist who can communicate on a level they can understand. Journalists in the electronic media use people they know are good reliable "talent".

On occasions I have been told by the less informed scientists, that using effective public relations to communicate science is prostituting the science itself. I have to tell you I am not a pimp, nor have I ever been, nor do I have any intention of becoming one. What I do is what all scientists should be doing, i.e., getting off your collective backsides and getting out into the community to let the general public know what your research is all about on a level that they can understand.

On arriving at the conference in Townsville I noticed some exquisite native birds and shrubs, the colours were quite dazzling. The sad fact is that in years to come, should my grandchildren visit the same area, there is little likelihood of them finding the same type of native birds and flora. Who is to blame? Every ecologist who has not communicated properly the results of their research. If the biodiversity of this country declines over the next twenty years, every ecologist at this conference is to blame.

You are totally responsible. I make no apologies for my statement, the apologies I leave to you. Why? Because you have not communicated the importance of your research — you have not informed the people who make the difference, and in spite of all your hard work, it has made no difference in the end.

[1]CSIRO, Division of Wildlife & Ecology, LMB 4, P.O. Midland, Western Australia, Australia 6056.

Pp. 46–49 *in* ECOLOGY FOR EVERYONE: COMMUNICATING ECOLOGY TO SCIENTISTS, THE PUBLIC AND THE POLITICIANS ed by R. T. Wills and R. J. Hobbs. Surrey Beatty & Sons, 1998.

Unless people know what you are doing, and the importance of your work, how are they to support it. From the Prime Minister to the man-in-the-street, everyone is affected by the loss of a species. People do care, and given the correct information will do something about it.

The benefits of good communication include the following:

1. Something effective is being done with your research.
2. You are being publicly accountable — remember your research is funded by tax dollars, AND some company donations.
3. Your stakeholders are getting that all important pat on the back for being astute enough to invest money in your project. They go back to their shareholders etc., and show them the benefits of their investment.

Do not think for a moment that your science is the most important outcome for the organization that is financing your research. There are very few philanthropists. The majority of companies and government departments know they gain from public recognition. The company (or product) gets credible publicity, is advertised in a more subtle way than straight advertising, and with newspapers or television reaches a vast audience. Why do you think Public Relations and Advertising companies charge such exorbitant rates?.

One project in the laboratory where I work has received over $250,000 of publicity, in five years. $50,000 a year is not bad for an investment of a part-time employee who works 15 hours a week for all projects on the research site, not just the one mentioned.

In my Division this is not exceptional, but quite normal. The mining company who has invested money in the project is obviously delighted. Therefore your first aim should be to ensure you have an effective communicator employed by your organization. Your second aim is to make the best of every opportunity this person provides for you.

GOOD COMMUNICATION SKILLS CAN BE LEARNED

Good communication = passion. To be a good communicator, you have to believe in your work, find it a challenge, exciting and be passionate about it. Those emotions already make you an effective communicator, all you have to do is learn a few techniques.

Public speaking

I cannot possibly cover all forms of communication, and so I will concentrate on general media tips. However, since seminars, symposiums, and general speaking engagements would take up more time than media for many people, they are extremely important so ensure that your skills are highly proficient in this area.

Nerves are always a problem, the thing to remember is that everyone feels nervous when speaking to an audience. Often the nerves you feel help to pump up your adrenaline, and give a fine edge to your performance. The secret is to control the nervous tension, make the butterflies in the pit of your stomach fly in formation, instead of any which way!

How do you overcome nervous tension? Well, there are ways and means!

The reverse breathing technique is very useful. Simple — any Mums will remember this from their time in labour. Come to think of it, this should bring back memories to Dads who were at the birth as well!

Stand up, and as you breath in your stomach should expand outwards, when you breath out, your stomach, should go in. It is the reverse of normal breathing, and by holding on to your stomach, and pushing in when you let your breath out, it helps to get into rhythm. Do this several times. The concentration on mastering the technique makes you forget about your nerves! Be careful not to hyperventilate!

Tension release is another useful skill. Clench your fists, and continue to tense every part of your body until even your toes are taut, then release the tension in one go. This only need to be done two or three times, to work effectively.

Standing before an audience is a very different form of communication to performing on electronic media, or talking to a journalist.

It is extremely important that you become good at it. Toastmasters and other organizations conduct excellent programmes that give you the techniques to perform effectively. A quick one or two-day programme will not give you sufficient practice and critiques of your performance to set in place all the techniques you require.

Media interviews

Your communicator has got you an interview with a journalist. So how do you prepare? The first step should be to make sure you are fully conversant with the topic, and you have a good idea of the type of questions you are going to be asked. If you do not know, find out, your communicator should have already made sure this is organized.

One way to ensure that you are asked the type of questions you want is to give as much well prepared, easy to read information to the journalist as you can. Make sure you only give them the information you want to talk about.

Talk to the journalist beforehand — a telephone conversation will help them get to know you and vice versa, plus you have got them thinking along the lines you want the interview to go. It is always important to get the journalist on side.

You can never ever be too prepared. Good planning and preparation = Good performance.

Being well prepared and well planned applies to any form of communication you carry out. Before you have an interview with a journalist, go through what you want to say over and over again, until you are sure that you are comfortable with the subject matter. Make notes of the salient points, and stories you can use for padding, if the interview goes on too long. Although you have fed the journalist as much information as you possibly can, NEVER and I mean NEVER trust the journalist to ask you all the questions you want. Always keep in mind the journalist is looking for a story and sometimes the facts can get in the way of that, so make sure you keep your answers clear, unambiguous and succinct.

If your mind goes blank, an easy way out of the difficulty is to say to the journalist, "Please could you repeat the question as the point you are making is not very clear".

Even if it was clear, it puts the journalist on the backfoot and gives you time to think of an appropriate answer. If the journalist is aggressive, a cool, calm response to their questions loses them respect with the audience, but gains a great deal of respect and the appearance of complete authority for you.

To avoid controversial questions, tell them bluntly but nicely:

1. They are looking for controversy where it does not exist.
2. Why ask the question when they are well aware you are not qualified to answer it.
3. What is the point they are trying to make?
4. You can always say the question is:
    (i) good
    (ii) interesting
    (iii) an intriguing one,
    (iv) however;
        (a) are asking the wrong person
        (b) the information is not available at this time.
        (c) When the type of information/data becomes available you would be delighted to discuss it further/at a later date.
        (d) the journalist has been misinformed and say why.
        (e) it would be unethical to answer the question as all relevant information is not available.
        (f) policy statements, as the journalist is aware, have to come from the appropriate Government Minister.

Never use the phrase, "What I meant to say was/is". If you meant to say it why didn't you? It also makes everything you have said so far completely irrelevant.

Also, never say "No comment", as it sounds like you have something to hide.

Never refuse an interview outright. Always sound disappointed, but give the journalist a good reason why you are not available. It leaves the option open for you to use the source when it is beneficial to you. Always keep in mind that you are the key to the interview, and you are the informed source. It will help you to control the interview.

Television

Practice answering questions in the mirror. You might feel foolish, but it does help you to see how you perform visually. One tip: lock the bedroom door. If you are discovered by a family member, it could cause unwanted social ridicule.

If visual aids, film or slides are needed, get them to the studio a day beforehand and let the producer know, but do not be offended if they do not have time to use them.

Make sure the clothes you are wearing are comfortable, there is nothing worse than sitting in clothes that feel like they are strangling you. Be careful of colour, red can "bleed", and stripes, especially bold ones, can waver and be irritating to the viewers. Smaller stripes and patterns can also waver. Ask advice from the producer. Try not to wear anything that distracts the viewer from the importance of what you are saying.

Scientists sometimes look like recycled hippies. If you want to get your message across as a serious scientist, gentlemen should wear a shirt, tie, decent trousers, and ladies a smart dress or suit; both may wear a jacket if appropriate for the climate. Never go on television looking like an unmade bed, because your credibility with the viewers will go down the toilet. If you are in the field, wear a T-shirt that promotes your stakeholder or your organization.

Unless you want to look like a two-day-old corpse let the make-up lady have her way with you. Her job is to make you look good, take advantage of this.

Practice your relaxation exercises, think of something calming. If it is morning, think of lunch — by that time it is all over — if afternoon, the evening meal!

Do not fret over your performance — just be yourself. Sit comfortably, and as you would normally sit, unless you are inclined to hang one leg over the side of the chair!! Control your body movements, flinging your arms around can cause ructions with the microphone.

Remember, the interviewer is interested in your research, you know your work and how good it is, get that enthusiasm, excitement and passion across to the audience.

Radio

Again follow the techniques used for television prior to the interview. Be prepared, only lack of preparation will cause you problems. Tape record your voice — play it back again and again, listen to the voice modulation, ensure you have enough expressiveness in your voice. Do not over do this, just enough to give yourself and idea of how you come across, get members of the family to help.

In the studio, remember those relaxation exercises. Ask the producer if the microphone will be in the way of your papers — if so get the producer to adjust the mike accordingly. Ask if you need headphones — if so ask the producer how to use them. Make sure they are adjusted to suit you and that they are completely comfortable. Make sure you are comfortably seated. Ask for a glass of water before you go into the studio — nervousness will make your throat feel dry. Open your file, take out the papers you need. Do not shuffle them on air because it sounds like a windstorm.

Do not under any circumstances:

- Tap your fingers or drum them on the desk, it sounds like a thunderstorm to listeners.
- Play with the microphone cord — to a listener it sounds like gale force winds.
- Shift or tap the microphone. It creates a cacophony of sound effects.

Telephone

Remember familiarity breeds contempt. Because you are comfortable talking on a telephone, you may be lulled into saying things you would not in a studio.

If you have just been contacted by a journalist about a story they want to do about your research, agree if you are happy to do the interview, but never do the interview immediately. Offer to call them back, if the matter is urgent in five minutes — longer if you require the time. Ask them about the questions they will be asking, and the type of story they want. If a radio interview, make sure you know how long before they want to talk to you. The five minutes or so you have taken can be used to gather the information you want to get across, and to relax you before the interview. Sit down and write all the salient points you wish to make during the interview. Make the points, brief, succinct and in order of importance.

Write in large letters, and place it in front of you, "I AM TALKING TO A JOURNALIST". Carry out relaxation techniques, then call the journalist back. The telephone is an all too familiar form of communication, you can easily forget you are being interviewed, and say things that could be regretted later. Keep focused and remember there is a journalist at the other end of the phone not a friend.

CONCLUSIONS

Good effective public relations and good communication does not prostitute science, it enhances it. It gets the information out into the community, and the community consists of people who can and are making a difference to our ecology. The community uses the knowledge it gains from you and does something constructive with it.

Whether Australia goes forward to the 21st Century with sound environmental policies and good management strategies is up to you, and depends on the way you, as environmental scientists, convey your message. Good science needs good communication!

CHAPTER TWELVE

Media tools and media traps: some suggestions for ecologists

MICHAEL C. CALVER[1]

Although some Australian ecologists are highly experienced in interacting with the media, most probably have little experience in preparing statements for the media, planning media exposure or responding to unsolicited media contacts. This chapter is a brief collection of ideas for dealing with some of the common problems or concerns in the area, which might otherwise escape the attention of busy researchers or academics. It is intended to complement the advice of public relations professionals and to encourage ecologists to share their skills and experiences in media presentation through the electronic and print bulletins of the professional societies and in professionally designed workshops.

> Interviewing experts can be a chore for questioner and listener alike. Experts often slip into their own technical language, sometimes deliberately to confuse the issue. I've often thought it would be good to superimpose sub-titles on television. The most tempting solution is Pythonesque — over-dubbing an interpreter's voice. "What the good professor is trying to say is . . ."
>
> Olle and Pullan (1992, p. 119)

> Slimy white things that crawl out of sewers.
>
> John Gorton on journalists, quoted on the front cover of *Tiffan* (1989)

> In some ways, too, the scientific method and the reporting method are simply at odds with each other.
>
> Fumento (1993, p. 338)

INTRODUCTION

ECOLOGISTS can learn easily to communicate effectively, persuasively and concisely using all common forms of print and electronic media. In this chapter you will see how, in an incredible 30 seconds, you can grab and keep the attention of readers, viewers or listeners, tell a compelling story, ask for what you want and get it! You will be able to make your point to colleagues, journalists, business executives, public servants and politicians and the whole range of the general public. In the long term you will even save time and improve your productivity. The steps and techniques to achieve this are simple and even fun.

Actually, the above paragraph is not my purpose in this chapter at all. It is a parody of the "30 second message" championed by Frank (1986) as a communications strategy for many walks of life. However, as you shall see argued with wit and conviction elsewhere in these proceedings (Kelly 1998), its central elements of interest, brevity and action are virtues in the communications media. The concept of a 30 second snatch on television or radio news may sit uneasily with scientists who are accustomed to longer discourse involving careful examination of methods, data and interpretations and who are instinctively suspicious of approaches that seem to value rhetoric over substance. However, if one accepts the view that communicating ecology via the media is vital, then it will be necessary to accept the constraints and work effectively within them. This means that except for those lucky few who have the opportunity of a feature in the press or a major interview on radio or television, we are stuck with the "30 second message". Fortunately, there are many sources, including the two given above, giving clear guides on constructing and presenting these vital snippets.

How, though, do you reach the position where you have the chance to deliver your message and what pitfalls should you beware of on the way? When I raised this point with the editors, their response was to invite me to research the topic and contribute this piece. Although my media experience is restricted to a few "letters to the editor", a handful of radio interviews and a couple of nervy appearances on television, for those in the same position there may be value in sharing what a novice finds important when entering this arena. It might also encourage colleagues with greater experience and insights to share these with

[1]Biological Sciences, Murdoch University, Murdoch, Western Australia, Australia 6150.

Pp. 50–56 *in* ECOLOGY FOR EVERYONE: COMMUNICATING ECOLOGY TO SCIENTISTS, THE PUBLIC AND THE POLITICIANS ed by R. T. Wills and R. J. Hobbs. Surrey Beatty & Sons, 1998.

other ecologists via the Ecological Society of Australia (ESA) electronic bulletin board. Contributions highlighting further specialist references would be especially valuable. The plan of the chapter is simple. It begins with a brief survey of the characteristics of print, radio and television media, overviews issues of organizational policy, confidentiality and ethics on the part of ecologists dealing with the media, outlines methods such as media releases, media events and direct contact with journalists for offering news to the media, summarizes presentation techniques appropriate to the different media, suggests how to respond to unsolicited contacts from the media and concludes with ideas for further debate/action by the ESA on these important topics.

LOCKING ON TO CHARACTERISTICS OF THE MEDIA AND OF JOURNALISTS

Both textbooks for journalists (e.g., Bell and van Leeuwen 1994; Clayton 1994) and "how to" handbooks for those seeking publicity through the media (e.g., Hailstone 1990) reiterate that news must sell papers or keep viewers or listeners tuned in. It may do this through sensationalism and scandal (which is probably not what most of us want for our science), or through informing and entertaining (which is closer to what many of us try to achieve). Rafe (1991, pp. 4–5) encapsulates both the selling concept and the presentational differences between the types of media in terms of conflict: "Reporters for newspapers and magazines look for data that support their concept of conflict and seek out reports and documents that can be excerpted and officials who can be asked questions that elicit meaty, quotable statements. Radio reporters try for the on-the-scenes quotes that bring immediacy to their coverage and sometimes use aggressive questioning techniques or loaded and leading questions as they press for responses. Television reporters also need immediacy, and they look for 30 second quotes that can be accompanied by visual drama — fires, the scene in front of the courthouse, and so on". While this is a very brief and possibly a jaundiced view, it does carry the important messages that effective media communications are short and compelling and tailored to the needs, potential and limitations of each medium used, while successful media communicators will very likely have to respond to conflict.

Another fact of life for many journalists is pressing deadlines (see, for example, Dunlevy 1980 for an account of the critical nature of media deadlines). When a journalist comes following a story lead the academic/scientist may want hours, even days to think about giving the details, while the journalist is probably facing a deadline that afternoon or at best that evening. Journalists tend to think in terms of minutes rather than hours. They also work on the pyramid model. The punchline (i.e., the main message, conclusion) is encapsulated in 25 words or less of the first paragraph and the story gradually unfolds from there, working on the principle that the least important facts are at the end where they can be conveniently cut by a sub-editor without, seemingly, harming the general accuracy and meaning of the story. Therefore, if you want to be heard, watched or read, you may need to be ready at very short notice with a brief but powerful message.

It surprised me to find that there are guides to good scientific writing that could actually help ecologists communicate with journalists under such pressures. Here are three handy hints that Brown *et al.* (1993) proposed for clarifying messages in scientific papers that seem very adaptable to communications with the media:

1. "Ask yourself 'if the people who are going to read this [scientific] paper [or, in our case, listen to this interview or read this story] remember only one point, what would I want that point to be?' Write that point . . . in *25 words or fewer.*" Brown *et al.* (1993, p. 184).
2. Pick three real people you know and would like to reach with your message (not categories such as "forest ecologists" or "wetlands biologists" but individuals such as aunt Agatha or cousin Lou), think of their needs and interests, and check that these match exactly with your main message. If they do not, revise the message.
3. Points 1 and 2 should prepare you for those precious 30 seconds on air or five lines in the press. If you have a little more media time or space, try preparing answers to each of the following questions in no more than 25 words per question. They are intended as a guide to writing scientific abstracts, but they also seem to fit the needs of media communications:
   (a) "what did you do?
   (b) why did you do it?
   (c) what happened?
   (d) what do the results mean?
   (e) what is your work good for?" (Brown *et al.* 1993, p. 186).

Specifically for ecological studies, one might add the further questions "what is the cost?" and "what are the risks?" since these seem to be common issues in ecological debates.

As ecologists we are all aware of the value of selecting appropriate journals for publishing our scientific work and tailoring our writing to their requirements, so perhaps similar thought could go into preparing and delivering our views in the media.

Caution: Ensure brain is engaged before putting mouth (or pen) in gear

It is probably prudent for any ecologist to consider organizational policy (including whether there are public relations colleagues who can help), if the issue is confidential (information that is commercial in confidence, for instance) and the ESAs own code of conduct for members participating in public debate (see the development of drafts in Kirkpatrick 1996a,b) before initiating any contact with the media. The first point seems almost too obvious to be worth stating, although I am aware of cases where people such as eager postgraduates or appointees new to an organization have disregarded it to their embarrassment or discomfort. Superficially, confidentiality seems a straight forward issue as well. However, outside obvious cases where you are constrained by a confidentiality agreement, you might want to think about the impact your views will have on other interested parties when they encounter them without warning in the media. Some mollification may be achieved and dialogue kept open in cases of conflict if others are notified in advance, for instance by giving them a copy of any planned media release. Sadly, one must sometimes be aware that the release of one's views may provoke some form of institutional retaliation (see the comprehensive documentation, including Australian cases, in Martin 1981, 1992 and 1993). Individuals and organizations must decide their own response in this situation. Finally, although the ESAs code of conduct for public debate is still only in draft form and will be binding only on members, its healthy advice that (i) opinions should be held honestly and their limitations acknowledged, (ii) reflections on the character and motives of other disputants are inappropriate and unacceptable, (iii) it is helpful to identify clearly areas of dispute and highlight the research or data that will resolve them, and (iv) that debate should be open, with no surreptitious pressure applied, could be borne in mind during preparation for any media release.

APPROACHING THE MEDIA

Newsworthy items can be circulated to the media via media releases, media events or direct contact with journalists. The first method is the most popular. In all cases, you are making the approach so you will need to fit the expectations of the editors and journalists if your news is to have a chance of being accepted.

Media releases

When reading a range of "how to" publications on writing press releases I was struck by the repeated emphasis on the importance of presentation. For example, one UK editor commented: "You can see the importance of presentation. Some stories will always be rejected because they do not fit the market. Some will always be accepted because they are important. With the bulk of stories — 71% in this example — professional presentation can make the difference between print and the wastepaper basket" (Peter Ransley, quoted in Jefkins 1968). Those fortunate enough to have public relations professionals employed in their organizations will no doubt receive excellent support in preparing a media release, but the following suggestions may be useful for those who have to be self-reliant. They are based on the approaches of Jefkins (1968), Mathews (1984) and Hailstone (1990).

The recommended sequence is:

1. The date and time the story may be released, which may be useful if you wish to give notice of a story but not release it for a short time. Hailstone (1990) advises that if you wish to publicise a forthcoming event you should give at least two days' notice, while for Australia Mathews (1984, p. 105) gives a detailed table of probable working hours of journalists and ideal contact times.
2. Subject of the story (given as a headline, which can be repeated on each page of the release).
3. Name of the organization in full, followed by the acronym.
4. Location of the organization (especially if it is different from the contact address for further information).
5. Advantages or newsworthiness of the subject or, as one author put it, start with a positive angle.
6. Applications of the subject, or why should anyone bother. Points 5 and 6 can often be run together in a single concise leading paragraph or even sentence, answering what Hailstone (1990) regarded as the essential questions of "*who, what, when, where, why* and *how*". As an example of how this can be done concisely and powerfully, he drew attention to the opening sentence of Hemingway's

"The Old Man and the Sea": "He was an old man who had fished alone in a skiff in the Gulf Stream and he had gone eighty-four days now without taking a fish". This answers all the essential questions except, perhaps, for why. For the skeptics, try comparing this approach to the five key questions for writing a good scientific abstract (Brown *et al.* 1993, quoted earlier in this chapter).

7. Further important details.
8. Source of further information. If possible, give an after hours contact as well as a work one, since journalists do not necessarily work your office hours, or even give two contacts to be truly sure.

All authors stress the value of attributing statements to individuals whose names and positions are stated clearly. Colourful, accurate quotes are useful if available, although one should beware of defamation. Mathews (1984, p. 74) quoted the example "pack of putrescent pansies" used by a Queensland MP to describe her parliamentary colleagues, but that was presumably said inside the House. Recommended length is absolutely no more than two pages, with just one preferred. To leave no doubt that the complete document has been received, each page should be numbered and the word "more" placed centre bottom of all but the last page, which should finish with "end" "###" or "XXX" centre bottom. Of course, for scientists it should be second nature to thoroughly check all facts and statements, especially names, dates, times and data.

The final press release might be posted or faxed to a wide range of media outlets including press agencies such as Australian Associated Press (AAP), or deliberately aimed at certain media, programmes or publications with a view to reaching a particular audience.

Media events

Jefkins (1968) uses this term collectively to embrace media conferences, when journalists gather to hear an announcement and ask questions; media receptions, which commonly begin with a reception/refreshments followed by a programme including events such as speakers, videos or demonstrations with opportunities for questions; and facility visits, which are invitations to tours, official openings and the like. Hailstone (1990) adds a further category of "media briefing" to cover situations where a story has already broken but there is a need to call journalists together to add more details. Some authors doubt the effectiveness of media events. Rafe (1991) claimed that they are rarely warranted and that one to one communication is best, while Mathews (1984) observed that cynical journalists may suspect an inverse relationship between the hospitality at a media reception and the newsworthiness of the story.

All categories are clearly only applicable to major stories and need meticulous organization, so they are best left to organizations with experienced public relations personnel. Detailed suggestions are given in Jefkins (1968), Mathews (1984), Hailstone (1990) and Rafe (1991).

Contacting journalists directly

Personal contact via telephone or even walking in off the street is possible (see Mathews 1984) but is hardly likely to endear one to a busy journalist. However, if you or your organization has dealt with particular journalists in the past then an informal first contact might be helpful in deciding how to handle your story.

PRESENTATIONS IN THE MEDIA

One possible outcome of a successful press release or direct contact with a journalist is an interview, either to provide background for a newspaper article or perhaps to be aired in whole or in part on radio or television. The subject of how to deal with an interview is large, so I will only cover briefly the topics of the rights of interviewees and preparing for interviews, with attention to some special features of interviewing for radio and television. Much greater detail can be found in specialist "how to" books. I found Rafe (1991) particularly clear and it also features a substantial bibliography, while Hailstone (1990) has point by point summary lists for preparation techniques.

Your rights

Jokes aside, journalists do have codes of professional ethics and reputable ones will follow these scrupulously during interview. Consequently, you should have little difficulty in gaining agreement with the following points. They are taken primarily from Rafe (1991, pp. 12–13), with material from other sources added where indicated.

1. You should know the name of the interviewer/s and the organization involved.
2. You can negotiate ground rules prior to the interview.
3. You should not be abused (although you may be asked tough questions!).
4. Your off the record comments should be kept confidential and not used in any context. However, there are pitfalls in

speaking off the record. While reputable journalists will never use information gained in this way, they may seek determinedly elsewhere until they find documentary evidence or someone else willing to say it (Clayton 1994). If this is your intention, that is fine. Otherwise, watch out.

5. You should be physically comfortable (chair, setting, etc.). You can complain if microphones, lights or other equipment are bothering you. Of course, you can do a lot to help yourself such as visiting the lavatory beforehand or not wearing heavy clothing to a television studio, where you may be hot under the lighting.
6. You can ask about the issues or subject of the interview in advance so you can be prepared. However, as Clayton (1994) points out, many interviewers are reluctant to release precise questions in advance as this may lead to a stilted interview without any sign of spontaneity.
7. You can ask the length of the interview in advance.
8. You can ask whether others apart from the interviewer will participate and for what purpose, or if you like ask to be accompanied by your own public relations representative.
9. You should be recorded only when advised, not during pre-show chats or commercial breaks.
10. You should have opportunity to answer without interruptions and to make your own points, provided of course you are brief and stick to the topic.

A thorny point that sometimes arises is whether or not you will be able to read the final story for a newspaper, or check the script of a pre-recorded radio interview before it goes to air. I did not find a source that advised journalists to do this and Clayton (1994) counselled against it, although she did recommend having factual accuracy checked by an independent authority. It seems important to appreciate that journalists view their pieces as their creative work and resent the tinkering of others, much as we might feel about interference with our own papers. They also have a low opinion, right or wrong, of our sensitivity to their deadlines: "Many boffin-types work in an environment where time does not exist. You send them your write-up and they could put it aside to attend to it 'later'. Then other papers get put on top of it and your piece sinks without trace" (Clayton 1994, p. 70). You might be successful by explaining that you would not read the piece for opinion or style but solely to check facts or technical information, or you could urge them to have the facts checked independently before printing.

Finally, it is very likely that your contribution will be edited substantially, prefaced with an introduction unlike what you expected, or otherwise placed in a context that you find uncomfortable. A good journalist will explain this to you before an interview, but you have no control over the process. Cop it sweet.

Preparing for interviews

Preparation for interviews is recommended widely and can include elaborate coaching exercises such as videoing and evaluating rehearsals (Rafe 1991) or more mundane but nevertheless valuable steps such as listing and rehearsing key points you would like to make, anticipating questions and planning answers for them, and devising a punchy concluding remark (Hailstone 1990). Generally, your time will be very limited. Commenting on the US media scene, Rafe (1991) noted that radio and television interviews used to last around 15 minutes but that this had now dropped to at most 10 minutes for radio and commonly two to four minutes for television, while the average time for a response had fallen from a minute to 25 seconds (close to the "30 second message" of Frank 1986). Consequently, Rafe recommended preparing two or three main points to make in an interview and to practice different ways of presenting them. The Australian scene is clearly similar, since the Australian media handbook Hailstone (1990) suggested planning no more than six points to make in a radio interview and only one for television. Whether you prefer KISS (keep it simple, stupid) or just SS (short and simple) the message is clear.

Describing the special features of radio, Barry Hill (quoted in Mathews 1984, p. 140) noted that: "When we listen to the speaking voice, there seems to be nothing to stop it entering our minds directly, with the same ease that a lover might whisper in our ear". While I suspect that Mr Hill has never listened to the broadcast of a football game, his comment does emphasize the value of voice modulation and delivery on radio and this skill can be practiced. With regard to television, Albert Mehrabian (quoted in Rafe 1991, p. 17) commented that 7% of a speaker's impact is content, 38% is voice and 55% is non-verbal such as dress, appearance and body language. While there is no space to detail the possibilities in each of these areas, it will clearly be helpful to follow carefully any relevant advice from your public relations department, or to consult the comprehensive suggestions in the publications quoted above. They also document types of interview questions with suggestions on how to deal with them. Entertaining transcripts of interviews in which some of these techniques are used are given in Olle and Pullan (1992) and Bell and van Leeuwen (1994). Further curly questions that were asked of a team of ecologists who studied the North American Northern Spotted Owl *Strix occidentalis caurina* are quoted in Shrader-Frechette and McCoy (1995, pp. 228–29). Although these were asked by politicians, one can imagine them being used in a media interview.

AMBUSHED

Unexpectedly finding a journalist at the other end of the telephone or standing in your office doorway can be a nerve-wracking experience (it was to me), but to put yourself back in charge remember that by agreeing to talk you are actually doing the journalist a favour. Clayton (1994, pp. 65–66) advises young journalists to repeat to themselves: "I need to interview them. I must remember that they do not need to be interviewed. All the benefits are for me . . . So I have a big persuasion job each time I approach someone for an appointment. I must not be belligerent or superior — I have no divine right to extract information from anyone". Even if your journalist is pressed by a deadline and has followed the rugged advice of Dunlevy (1980) that therefore it is OK to barge in, you remain the one with the favour to give. Rafe (1991) recommends responding politely but assertively. If one or more journalists arrive in person, you can introduce yourself and shake hands, which may put them off their stride, before talking to the least aggressive or quietest person. If you cannot talk then say so and give a reason, but explain how they should contact you to see if an appointment is possible. Otherwise, state clearly that you only have a limited time for questions and invite the quietest person present to begin. When the unexpected contact is by telephone, explain politely that you cannot talk then but will be happy to call back at a later time. Ask for a brief account of what s/he wants to discuss and end the call. Persistent journalists may not be put off easily (Clayton 1994 gives an entertaining list of arguments and ruses for them to try), but by sticking to the call-back approach you gain time to marshal your thoughts, perhaps check with others that you are permitted to discuss these issues at all and you can confirm the bona-fides of the caller. I have not tried this approach myself, but it certainly sounds useful.

WHERE DO WE GO FROM HERE?

Of course, I claim no originality on my behalf for the ideas expressed here, nor have I tried many of them in practice. Perhaps, though, the exercise of collecting the points in a place accessible readily to ecologists might guide those contemplating offering their views to the media, or suddenly finding themselves in the midst of a public controversy. I was also surprised to find such strong parallels between some of the advice for better scientific writing and the special features of writing/presenting for the media, so media communication may not be such a foreign exercise after all. Some of us may also be able to help ourselves by encouraging our employers to provide workshops in media relation skills as part of staff development training. Media training sessions in which people practice before microphones or cameras under the guidance of experienced journalists might be especially helpful. As ESA members, we could also share experiences and ideas about this area in the both the print and electronic bulletin pages, while skilled and experienced members might contemplate organising a skills workshop in media relations to run alongside one of our conferences. I would certainly hope to attend.

ACKNOWLEDGEMENTS

I am grateful to Liz Wood for sharing insights from her experience in practicing and teaching journalism. Russell Elsegood responded to the first draft with some excellent suggestions which strengthened the chapter greatly. Many thanks to you both.

REFERENCES

Bell, P. and van Leeuwen, T., 1994. The media interview: confession, contest, conversation. University of New South Wales Press, Kensington.

Brown, R. F., Pressland, A. J. and Rogers, D. J., 1993. Righting scientific writing: Focus on your main message. *Aust. Rangel. J.* **15:** 183–89.

Clayton, J., 1994. Interviewing for journalists. Judy Piatkus Publishers Ltd, London.

Dunlevy, M., 1980. Interviewing techniques for newspapers. Deakin University Press, Deakin, Victoria.

Frank, M. O., 1986. How to get your point across in 30 seconds — or less. Corgi Books, London.

Fumento, M., 1993. Science under siege: balancing technology and the environment. William Morrow and Company Inc., New York.

Hailstone, R., 1990. The Australian media handbook. Information Australia, Melbourne.

Jefkins, J., 1968. Press relations practice. Intertext Books, London.

Kelly, K., 1998. Ecology TV. Pp. 42–45 *in* Ecology for Everyone: Communicating Ecology to Scientists, the Public and the Politicians ed by R. T. Wills and R. J. Hobbs. Surrey Beatty & Sons, Chipping Norton.

Kirkpatrick, J., 1996a. President's report. *Bull. Ecol. Soc. Aust. Inc.* **26**(2): 38–40.

Kirkpatrick, J., 1996b. President's report. *Bull. Ecol. Soc. Aust. Inc.* **26**(3): 84–85.

Martin, B., 1981. The scientific straightjacket: The power structure of science and the suppression of environmental scholarship. *The Ecologist* **11**(1): 33–43.

Martin, B., 1992. Intellectual suppression: why environmental scientists are afraid to speak out. *Hab. Aust.* **20**(3): 11–14.

Martin, B., 1993. Stamping out dissent. *Newsweek* **April 26:** 1993.

Mathews, I., 1984. How to use the media in Australia. The Dominion Press, Hedges and Bell, Victoria.

Olle, A. and Pullan, R., 1992. On interviewing. Australian Broadcasting Corporation, Sydney.

Rafe, S. C., 1991. Mastering the news media interview: How to succeed at television, radio and print interviews. Harper Business, New York.

Shrader-Frechette, K. S. and McCoy, E. D., 1995. Method in ecology: Strategies for conservation. Cambridge University Press, Cambridge, UK.

Tiffan, R., 1989. News and power. Allen and Unwin, Sydney.

CHAPTER THIRTEEN

Communicating with a poster presentation

ROBERT J. WHELAN[1]

THE POSTER — AN UNFAMILIAR MEDIUM?

POSTER presentations are becoming more common at scientific conferences, especially when the number of people wishing to make presentations is large. A decade ago, it was possible for all presenters to give a talk in the "Open Forum" section of the two- or three-day conferences of the Ecological Society of Australia. Now it would take four or five days of talks, even with concurrent sessions, to permit each presenter to give a talk.

In other, larger societies, in Australia and overseas, posters have long been a significant part of each conference. For example, at the annual meeting of the Australian Society for Microbiology, about half of the contributions are poster presentations (M. Walker, pers. comm.), and posters feature prominently at national and international biochemistry conferences (M. Wilson, pers. comm.). In some societies, posters are the predominant form of communication, especially for research students and early-career researchers.

Big conferences are also the places to see posters put to other uses. There are often sales displays by book publishers and by scientific supply companies. These companies base their displays on poster presentations. At Open Days of large organizations, at public lectures, at the public exhibition of development proposals for Local Government, and in many other situations, posters are a common means of communication.

Given the increasing use of posters as a means of communication within scientific conferences, and the widespread use of communication by poster in many other arenas, graduates should surely be equipped with skills in good communication via this medium. Some scientific societies have emphasized the importance of posters by awarding poster prizes. The Ecological Society of Australia, for example, awards a prize for the best poster by a student, at its annual conference. There appears to be little explicit training in preparing a good poster, in many University courses.

There may be several reasons for this. Posters are an unfamiliar medium for most teachers. We educators were trained (if we were lucky!) only in the forms of communication that dominated at the time we received our training — scientific writing, lecture presentations and short conference talks and seminars. Much has been written about what constitutes a good talk and how to give one (see (i) the wonderful summary, called "A Seminar on Seminars" by Prof. Reg Moir, *in* Lindsay 1995; (ii) articles by Janzen 1980, and Pickett *et al.* 1991, and (iii) the chapter on "Giving an Oral Presentation" in Jones *et al.* 1994).

We do not have such a clear idea of what makes a good poster, and there is much less consistency in the view of judges as to what criteria should be used in assessing poster "quality". There are few published discussions about or guides to poster presentations (but see Chapter 55 in Jones *et al.* 1994). This problem has been highlighted recently by the difficulty of judging the Australian Flora Foundation prize for the best student poster **or** spoken paper at the Ecological Society of Australia conference. Here, the quality of posters has to be judged against that of spoken papers.

In this chapter, I present some ideas about good poster presentations. They are based on discussions with a range of researchers and postgraduate research students. Any assessment of what constitutes a good poster must start with an statement of what one is trying to achieve in a poster — what is to be communicated? I then discuss features of good posters, based on experience attempting to judge posters for student prizes and consequent discussions with postgraduate students (see Whelan 1996). I hope that this chapter will stimulate more discussion and lead to further development of good poster presentation skills by students and researchers.

[1]Department of Biological Sciences, University of Wollongong, Wollongong, New South Wales, Australia 2522.

Pp. 57–60 *in* ECOLOGY FOR EVERYONE: COMMUNICATING ECOLOGY TO SCIENTISTS, THE PUBLIC AND THE POLITICIANS ed by R. T. Wills and R. J. Hobbs. Surrey Beatty & Sons, 1998.

WHAT IS A POSTER FOR?

Like a talk, a poster presentation can be expected to fulfill a number of functions:

- Advertise the research and the researcher;
- Convey a general message or conclusion;
- Communicate ideas, models, concepts, hypotheses, solutions to technical problems;
- Provoke discussion;
- Provide a more detailed back-up to a talk.

Not every poster would have all of these aims, and perhaps there are more than are listed here. However, whatever the aims for a particular poster, it is clear that masses of data and masses of text presented in the space of about 1 m^2 will compromise the aims of advertising the research and researcher and of conveying a general message.

CHALLENGES OF THE POSTER MEDIUM

I perceive the main challenges of effective communication by poster to be (i) finding the right balance between detail and impact (good advertising) and (ii) dealing with an audience whose reactions cannot be detected or respond to immediately. Many of the people who view a poster will do so when the author is not present, or is busy talking to someone else.

It is tempting to deal with the latter problem, which requires that the material in the poster is self-contained, by piling lots of detail into the text and figures. As mentioned above, this is incompatible with having an impact and advertising the work. Among tens to hundreds of other posters in the room, the challenge is to:–

- attract people over to **your** poster;
- make them want to read it;
- make it easy for them to follow the argument;
- highlight the main messages and make them memorable;
- convince them that the messages are based on work of substance.

In addition to these challenges, there is a host of (potential) technical problems:

- How big will the poster board be (0. 9 m^2 to about 2 m^2)?
- What will the alignment be (0.9 m wide and all the way to the floor or 1.5 m long and only 0. 9 m tall)?
- What will the material be (corkboard, hessian-covered pin board, cloth, brick wall)?
- What facilities will be available in the poster hall (powerpoint nearby, pins, bluetack, velcro, masking tape)?

FEATURES TO AIM FOR

Grabbing attention

The title and the overall "look" of a poster contribute to its attractiveness to the room full of people. The title has to have few words and also be in a large font size — and it should be a catchy phrase. Dan Janzen is surely the leader in composing catchy titles . . . e.g., "How to be a fig" (Janzen 1979) and "Mice, big mammals and seeds: it matters who defecates what where" (Janzen 1986), and there are numerous other models in the literature. Jones *et al.* (1994) argued that a poster title should be no more than eight words long and in letters >4 cm tall.

The overall layout of the poster contributes to its attractiveness from a distance, well before any words or other detail can be deciphered. Jones *et al.* (1994) presented a nice comparison of three different layouts which makes this point (see Fig. 1). Good use of colour also contributes to the attractiveness of the poster, but poor use of colour can surely drive people away!

Making them want to read it

Just as the overall layout of a poster contributes to its attractiveness, from a distance, so it contributes to the reader's desire to read on — beyond the title. The spatial layout needs to be interesting and pleasing to the eye. One way of achieving this is to amass the illustrative material first (pictures, diagrams, figure), arrange this material in an interesting and logical sequence, and build the text around them (see Fig. 1c).

The subdivision that this approach creates can have several benefits. First, it may constrain the amount of detailed text that can be written for each component of the poster. Second, it will allow the poster to be constructed in modules, which is always a good idea when the size and shape of the poster board available are not certain in advance. Third, it gives an opportunity to use coloured backing card (or some other technique) to produce a visual separation between the sections.

The amount of text, size and style of the font and the way the text is arranged can all influence the desire of the viewer to start reading (and to continue, once started). Jones *et al.* (1994) argued that there should be fewer then 500 words of text on a poster. Journalism provides the hints on how this should be

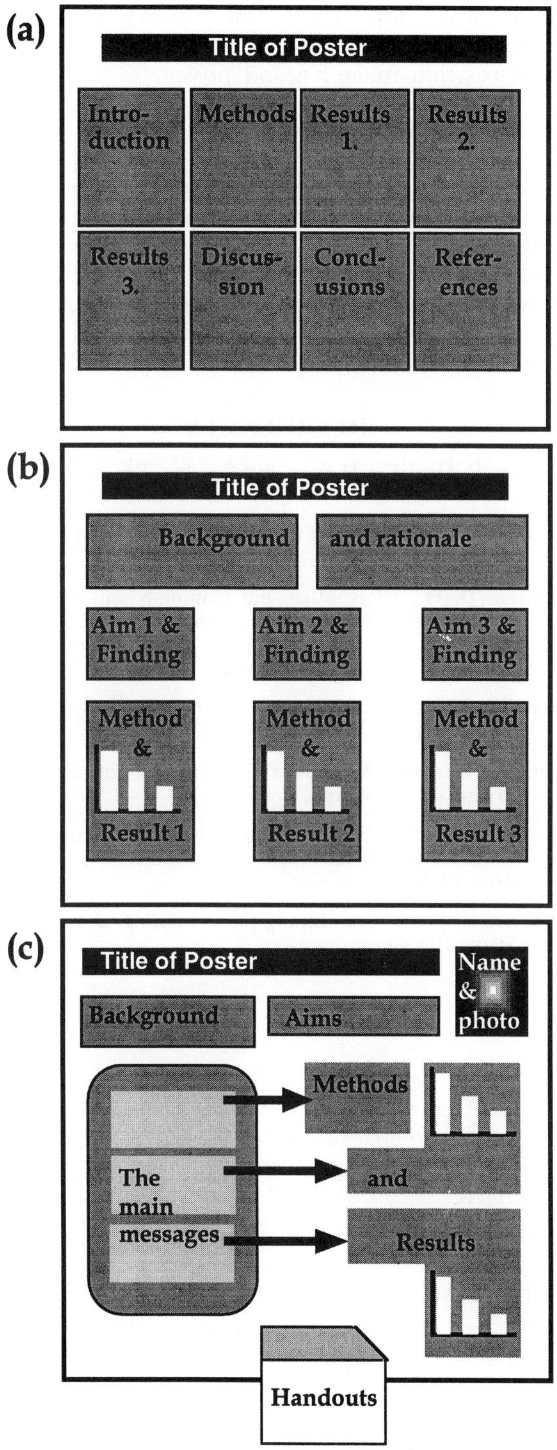

Fig. 1. Three contrasting poster designs (see also Jones *et al.* 1994).
(a) An uninspiring layout with each conventional section on a separate panel. Conclusions (main messages) at the end, not in a power position.
(b) A more appealing arrangement of panels, with combined methods and results arranged around pictorial material. Findings (main messages) in the power position of the poster.
(c) An example of a more innovative layout, focusing on the conclusions and linking each with its respective study (method and results). Note the photo and address of the author, and the envelope for handouts at the bottom.

arranged: (i) using short, concise sentences; (ii) using clear headlines and subheadings; (iii) avoiding text all in capitals, or in italics, or in outline or shadow font. The font size should be such that the text can be read easily from 1 m or more away — 18 to 24 point is appropriate and will help confine the poster to 500 words!

Making it easy to follow

Creative arrangement of the modules in a poster can provide an intuitive "map", with the attention of the reader directed from one section to the next. Otherwise, there is a need to label each module clearly, in some way, so that the various bits are read in the correct sequence.

Imagination provides the only limit to the ways in which these sorts of links can be contrived. A most effective example of this was a prize-winning student poster on greening disease of citrus, which was constructed in the shape of a citrus tree canopy. The Background and Rationale for the study were presented in the trunk, several hypotheses were presented in the major limbs, and for each of these there was a Methods branch, a Results branch and a Conclusion branch.

THE MAIN MESSAGES

One of the problems with a poster is the shape of the space available. The conclusion will contain the main "take-home" message of the poster and the conclusion logically comes at the end. Typically this would place the main message low down . . . yet the "power" position on the board is at eye level. This is, after all, the level at which supermarkets place the goods they are keenest to sell!

A challenge in designing a poster layout is therefore to find a way to place the main messages, or conclusions, in a power position on the board, or else to find another way to highlight them. They should be stated briefly and clearly.

CONTENT

Five hundred words is not much to play with, when there is so much to say. This constraint will obviously limit the amount of content which can be presented in a poster. However, it is better to have the viewer read and digest 100% (or even 60%) of 500 words than to abandon reading after the first 5% of a 5 000 word poster.

Based on the principle that "a picture is worth a thousand words", well-designed (i.e., self-contained) figures, photographs or tables can provide one way to present detail without adding much text. Nevertheless, there will always be much more to present than can be included in the poster. Furthermore, relatively few of the people who will view a poster would actually desire more detail. How can one cater for these few?

One useful strategy is to provide additional material as a handout, contained in a pocket on the poster board. A handout can include more detail on the methodology, more detailed results, some detail about statistical analyses, and some references to published work. In addition, a handout can include the author's name and contact details.

FOLLOWING UP THE POSTER

It is important to make it easy for readers of a poster aware of who the author is. Although most conferences have a poster session, during which the authors stand at their posters, not all people interested in a poster will necessarily see it at this time, and many will see it when the author is busy talking. A photograph is therefore very important — somewhere up near the name and title. The interested reader can therefore recognize the author at a tea break and follow up on the poster then.

CONCLUSION

There are few guidelines for preparing effective poster presentations. Given the increasing importance of posters in many arenas, more discussion is needed on the features that make a good poster.

ACKNOWLEDGEMENTS

The ideas presented here were stimulated and firmed-up in discussions with my research students, colleagues in the Department of Biological Sciences, University of Wollongong, and with fellow judges of talks and posters at many an Ecological Society of Australia conference.

REFERENCES

Janzen, D. H., 1979. How to be a fig. *Ann. Rev. Ecol. System.* **10:** 13–52.

Janzen, D. H., 1980. Plea from a symposium goer. *Bull. Ecol. Soc. Amer.* **61**(3): 170–71.

Janzen, D. H., 1986. Mice, big mammals and seeds: it matters who defecates what where. Pp. 251–71 *in* Frugivores and Seed Dispersal ed by A. Estrada and T. H. Fleming. Junk, Dordrecht.

Jones, A., Reed, R. and Weyers, J., 1994. Practical Skills in Biology. Longman, Essex.

Lindsay, D., 1995. A Guide to Scientific Writing. 2nd Ed. Longman, Melbourne.

Pickett, S. T. A., Hall, B. and Pace, M., 1991. Strategy and checklist for effective scientific talks. *Bull. Ecol. Soc. Amer.* **72**(1): 8–12.

Whelan, R. J., 1996. A winning poster? *Bull. Ecol. Soc. Aust.* **26**(3): 103–04.

CHAPTER FOURTEEN

Teaching communication in undergraduate biology

ERIC J. DORFMAN[1] and CHARLOTTE E. TAYLOR[1]

INTRODUCTION

THIS volume recognizes the need for ecologists (and scientists in general) to be better communicators. Skillful communication is vital in linking the public (and its opinions) with the intricacies of science and conservation, and many established scientists are now taking the time to acquire these skills. However, there is also an obvious need to address communication at an earlier stage in a science career. Recher (1992, p. 22) is particularly scathing in his appraisal of Australian science graduates:

> "They enter university with high school level English and verbal skills, and they receive their Ph.D with about the same level of expression. Along the way they have acquired the language of the field of their speciality and to survive (pass) they use jargon instead of the simple English they are incapable of writing."

Teaching individuals to communicate during the process of becoming professional scientists is an obvious pro-active step which is often overlooked, (Clanchy and Ballard 1995). However, there are distinct advantages in combining the teaching of communication skills together with scientific knowledge. Students who are made aware of the relevance of each, at an early stage in their professional education, will develop an understanding of the connection between the two. The interplay between science and communication can thus enhance the learning process overall.

Our aims in this chapter are to highlight the potential of the tertiary degree programme as a tool for imparting communication skills, and to present examples of a few of the many ways of accomplishing this within the context of undergraduate biology. We present examples of initiatives introduced into teaching programmes in the School of Biological Sciences at the University of Sydney. We have also included some examples of similar exercises employed in other subjects and universities in Australia.

INITIATIVES IN TEACHING COMMUNICATION SKILLS

A case study from the School of Biological Sciences, the University of Sydney

Teaching communication as an integral part of a biology course, was formalized after receipt of a CAUT grant (Drury and Taylor 1995) to develop a writing programme within the first year course. This course was targeted because students entering university need to be aware of the importance and relevance of communicating from the outset. The progression from the first year writing programme is now being addressed by developing links with individual initiatives in the second year courses. New writing components are also being developed for graduate courses in biology. A similar emphasis on communication skills is still required in other senior and honours courses since there must be an acknowledged continuity throughout the degree, if these innovations are to be completely successful.

Written skills

The most common mode of communication employed by scientists is still the written word, whether it be in the form of a formal refereed journal article, an expert or government report, discussions in the media, or interactions via email. Emig (in Nightingale 1988, p. 270) states that "writing leads one to integrate material, allows review and re-evaluation, helps form connections, and is active and dictated by ones own patterns of thinking and doing." However, writing is rarely taught explicitly within science courses. Most science co-ordinators assume either that students will acquire these skills elsewhere, or will absorb them through "osmosis". This attitude obviously devalues the writing process in students' eyes, and overlooks the acknowledged link between students' ability to communicate and their understanding of the subject (Penrose 1992; Ackerman 1993).

---

[1]School of Biological Sciences, A08, University of Sydney, Australia 2006.

Pp. 61–64 *in* ECOLOGY FOR EVERYONE: COMMUNICATING ECOLOGY TO SCIENTISTS, THE PUBLIC AND THE POLITICIANS ed by R. T. Wills and R. J. Hobbs. Surrey Beatty & Sons, 1998.

The "Writing in Biology" programme (Drury and Taylor 1995; Taylor and Drury 1996) has been specifically developed to be an integral part of the first year biology course. We are aware that writing needs to be taught in context, and encourage students to practice and improve their writing using relevant biological examples. We try to create tasks which mirror as much as possible "scientific writing", although this often proves difficult at such an introductory level, since many students have difficulty relating their course activities in first year to future work they may undertake as scientists. We create self-help exercises, notes and manuals which allow students to follow a progressive development in writing skills through different stages of the course. Small group activities are used in which examples of "good" and "poor" writing are discussed and the criteria for assessing good writing are developed. Students are then given a number of individual and group writing tasks followed by group and teacher feedback using the selected assessment criteria.

Student feedback has shown that the programme meets the needs of the wide range of students in the course. Individual students prefer different areas of the writing programme, but the following seem to be particularly useful: models of "good" and "problem" writing, the self-help materials (i.e., notes on "How to Write a Report") and the range of feedback materials. We have now produced a self-help manual on the transition into second year with particular emphasis on communication skills and building on first year experiences.

Visual skills

The two most frequent methods by which scientists communicate with their peers (i.e., journal articles and conferences) both require the use of visual material (e.g., graphs, diagrams and drawings) to summarize results. The goal within the School of Biological Sciences is to give students an opportunity to communicate their scientific knowledge visually through scientific illustration and the creation of posters.

First year students are introduced to the conventions for drawing in science and, in first semester of second year, students have the option of attending a formal course on scientific illustration. The course runs over two evenings, one concentrating on theory and the other on practice. Students are introduced to the concept of illustration as communication and as art, using examples from the early masters and modern artists. They are also introduced to techniques of form and proportion, the use of different materials (such as ink and pencil), and shading using stippling, hatching and smudging. Interspersed throughout the lecture, students are given the opportunity to draw specimens from the School's museum collection. Additional help in illustration is provided during formal laboratory classes and on a intensive three-day field excursion. Students who have availed themselves of these opportunities, frequently show remarkable progress.

In addition, students within first and second year courses are asked to construct scientific posters. In first year, students work in small groups to create a poster and a short talk based on a topic relevant to the course. They are provided with information on "what makes a good poster" and discuss the assessment criteria before presenting their poster for assessment and feedback by students and teachers.

In second year, students are required to construct a poster on invertebrates, either on diversity within a taxon, or as a summary of results of the field excursion. To give the students an idea of the format and the quality expected, a sample poster on the diversity of an order of birds (Pelecaniformes) is exhibited. The poster is annotated with information about formatting, style and content. Students are asked to answer questions which make them think about the layout and effectiveness of the poster. This is considered pre-work for a poster exercise, conducted during a discussion group.

Discussion groups are divided into two subsets of 5–10 students. Each sub-set is provided with a large laminated board and a wide selection of material on small laminated sheets, relating to the biology of terrestrial fish (mudskippers) and given one of two briefs: to create either a poster for the general public, or one for a scientific audience. Students are able to choose from multiple titles, colour photographs, charts, cartoons and blocks of text which vary in font size, length and level of detail. Students spend approximately half an hour choosing appropriate material, laying it out and affixing it with moldable adhesive. After this, each set of students displays its work and is asked, in a general discussion, to justify choice of material and layout. Students who have undergone this training subsequently produce posters which are generally substantially better than those who have not.

Student essays and reports often include summary charts and graphs which are inadequate due to insufficient knowledge of presenting such information. A knowledge of the conventions of graphical presentation is important as it also permits critical interpretation of the results of others. While the increase in use of graphical techniques for examination of data (e.g., multidimensional scaling), has also meant that images are increasingly involved with the generation of results, not merely their interpretation, a misunderstanding of such techniques can lead to inappropriate conclusions (Dorfman and Hochuli 1995).

Oral skills

The public's concern over the state of the environment has been growing steadily for the last three decades. Discussion of contentious environmental issues such as acid rain or "scientific" whaling present a challenge, because scientists cannot rely upon universally agreed codes of ethics or an appropriate body of laws as a basis for argument (Stephen 1992). A facility with language is thus essential in communicating the inherent complexity of these topics to non-scientists in a way that is clear enough to be understood and acted upon (Hutchings 1990). Biologists have been criticized for an inability (or an unwillingness) to communicate in such a way that they can be understood by the media (Recher 1992), yet it is frequently public opinion that is the deciding factor in many important issues. Clearly, training in the ability to argue a case lucidly should be included in any science curriculum.

Oral presentations, debate and group interaction are all important vehicles for promoting spoken communication skills. First year students give a short talk based on information in their poster on skeletons. This talk is not obligatory, although we have found that most participate as part of their contribution to the group effort. Feedback and discussion after these talks usually focuses on solving problems students have experienced, (e.g., being nervous or reading from notes). Students are also encouraged to take note of tactics used by "good" presenters among their lecturers. A video on presentation skills is being created in the school to cover all aspects of giving talks and making posters. It is expected that this will be used as self-help material or as part of structured discussions with student groups.

Students in second year are given an opportunity to hone their skills in bi-weekly discussion groups, in which they discuss journal articles on topics set by course instructors. There are approximately 15 students per group and topics range from theoretical concepts such as tautology and animal consciousness to more applied topics such as conservation methods and taxonomy. For a given topic, students read between three and five papers, which have been chosen for their controversial nature. Although there is no formal teaching of debating methods, discussions are usually animated and often heated, which obliges students to articulate their thoughts carefully. Discussion groups are habitually the part of the course which receive the highest praise in student evaluations.

EXAMPLES OF OTHER INITIATIVES

In the Department of Biochemistry at the University of Sydney, changes have been made in the teaching focus, to incorporate a number of generic skills (Denyer, pers. comm.). Communication skills are recognized as vitally important in training biochemists and emphasis is placed on the development of good writing skills in association with data collection and report writing. Feedback is acknowledged as a critical element in the progressive acquisition, and honing of communication skills; and students regularly review each others' work. This requires students to analyse and criticize other work in the light of how such comments can be applied to improving their own work.

A new approach to teaching has been introduced in the Professional Engineering course at the University of Sydney by Baillie *et al.* (1996), where a range of skills are practiced during the teaching and learning of the appropriate specialist knowledge. In this course, students work in small groups for most exercises, including preparation of reports on projects and practical applications. Students have a number of discussions on the concept of "good" writing and oral skills (within their discipline) and are involved in the development of appropriate assessment procedures. Communication is constantly discussed in terms of the "audience" and the concept of "getting the message across".

Kessell (1996), at the Department of GIS, Curtin University of Technology, has targeted the lack of written, oral and interpersonal communication skills, as well as a range of other "missing items" in Australian science degrees. He has developed courses in first year which include expectations of demonstrated competence in micro-computing, keyboard and word processing skills including touch-typing. The course also emphasizes spoken and written communication and the importance of being aware of the target audiences. A third year course in "Technical Writing and Project Preparation" gives more intensive practice in editing, "saying what you mean, stopping the redundancy, dropping the jargon", and writing and speaking to the level of the audience. All students learn the use of proper proof reading marks and use them throughout the unit. Students have seven or eight written assignments and have to give oral presentations. They also have to prepare a written resume and create a job application letter for a given job advertisement. This is followed by presentation to a panel of professionals who interview students in front of the class, an experience which "most students rate . . . as the most terrifying but also the most useful and valuable component of the unit" (Kessell 1996, p. 366).

THE FUTURE . . .

Universities now recognize the importance of teaching "generic skills" (e.g., communication and critical thinking), in addition to subject-specific ones, and competence in these areas is expected of graduates, both by university teachers and by prospective employers (Kemp and Seagraves 1995). Kessel (1996) argues that students who have mastered these skills should expect to be more in demand as professionals, and will be better equipped to continue developing and adapting their skills as flexible and adept "life-long learners" (Candy *et al.* 1994).

Australian ecology students are given a strong education in their field, but unless they can communicate their findings cogently they will "find it difficult to relate their work to the human community or even to understand that others might be interested in what they are doing" (Recher 1992, p. 23; see also Kessell 1996). It may be argued that ongoing economic constraints within universities will impose limits on what can be included in undergraduate courses. Regardless, if we hope to produce truly capable scientists, we as instructors must re-evaluate the philosophy of Australian ecology courses, and make a place for communication skills among the disciplines of biology, mathematics and statistics, which are already integral parts of the degree.

ACKNOWLEDGEMENTS

We would like to thank David Patterson, Dieter Hochuli, Helen Drury and Shaharin Yussof for discussions on the topic and for reading drafts of this chapter. We would also like to thank the reviewers for their comments and the ESA editors and symposium organizers.

REFERENCES

Ackerman, J. M., 1993. The Promise of Writing to learn *Writt. Comm.* **10**(3): 334–69.

Baillie, C., McHugh, P. and Davies, W., 1996. Professional Engineering — Teaching for better learning. Committee for the Advancement of University Teaching, Canberra.

Candy, P. C., Crebert, G. and O'Leary, J., 1994. Developing lifelong learners through undergraduate education Commissioned report No. 28 National Board of Employment, Education and Training, AGPS.

Clanchy, J. and Ballard, B., 1995. Generic skills in the context of higher education. *High. Educ. Res. Devel.* **14**(2): 155–66.

Dorfman, E. J. and Hochuli, D. F., 1995. A few thoughts on ordination and hypothesis testing. *Bull. Ecol. Soc. Aust.* **25**(4): 164–65.

Drury, H. and Taylor, C. E., 1995. Teaching writing skills in the science curriculum — final report for a 1994 National Teaching Development Grant. Committee for the Advancement of University Teaching, Canberra.

Hutchings, P., 1990. The dilemma facing biologists today. Pp. 71–79 *in* Zoology in Court ed by D. Lunney. Royal Zoological Society of NSW, Mosman.

Kemp, I. J. and Seagraves, L., 1995. Transferable skills — can higher education deliver? *Stud. High. Educ.* **20**(3): 315–28.

Kessell, S., 1996. Addressing the flaws in Intensive Science Degrees. *Res. Devel. High. Educ.* **19**: 364–70.

Nightingale, P., 1988. Understanding Processes and problems in student writing. *Stud. High. Educ.* **13**(3): 263–85.

Penrose, A. M., 1992. To Write or not to Write: effects of task and task interpretation on learning through writing. *Writt. Comm.* **9**(4): 465–99.

Recher, H. F., 1992. Simple journalists or simple scientists?: are environmental issues too complex for the media? *Aust. Zool.* **28**(1–4): 19–23.

Stephen, N., 1992. Environmental dispute resolution. *Aust. Zool.* **28**(1–4): 10–15.

Taylor, C. E. and Drury, H., 1996. Teaching writing skills in a first year biology course. *In* Different Approaches: theory and practice in higher education. Proceedings of the annual conference of the Higher Education Research and Development Society of Australasia ed by S. Leong and D. Kirkpatrick. **19**: 864–69.

CHAPTER FIFTEEN

Communicating globally significant conservation problems: Bolivia as a case study

KATHARINE J. M. DICKINSON[1]

INTRODUCTION

TODAY'S human population is placing ever-increasing pressure on the world's natural resources (Ehrlich and Ehrlich 1990). Communicating to the public the necessity to conserve biodiversity is a major challenge particularly given the rapidity of species loss and ecosystem change. On a global scale, the quantification of biodiversity remains rather vague (Dobson 1996). Furthermore, the location and extent of protected areas required to conserve biodiversity is a major source of debate (Daily and Ehrlich 1994). Wilson (1992) considers that even maintenance of existing legally protected areas (only 4.3% of the earth's land surface) is insufficient to save the species within them. He recommends that reserves should expand from 4.3% to 10% of the land surface and at the least, protected areas should include as many of the undisturbed habitats as possible with priority given to the world's biologically rich hot spots.

In general terms, we do know where some of the hot spots occur; often they coincide with poorly explored habitats such as the deep ocean floor and tropical rainforest ecosystems. In these cases, the lack of knowledge, often combined with ignorance of the significance of such sites, may translate to a lack of conservation advocacy and communication of what must be done to protect them. In such a situation, political, social and economic pressures can present a real threat to the protection of biodiversity unless there is a groundswell of public concern. Experience shows that such concern can only come from informed citizens and pressure groups even within our own countries. We can take examples from such cases as the successful Save Lake Manapouri campaign in New Zealand (Peat 1994) and the Franklin-Lower Gordon rivers campaign in Tasmania both of which eventually culminated in establishment of a World Heritage Area (Hutching and Potton 1987; Smith and Banks 1993).

When it comes to influencing practices which involve another nation's natural resources especially in the developing world, communication, education and innovative solutions involving monetary compensatory action are often necessary (Wilson 1992). Use of natural resources is understandably seen by developing nations as the means to economic advancement (Perrings *et al.* 1994) and their conservation may not be a priority — a road already followed by developed nations (World Commission on Environment and Development 1987; Shiva 1988).

Tropical ecosystems harbour a vast proportion of global biodiversity — particularly rich areas are found in Amazonia, Indonesia, Malaysia and West Africa but some tropical areas have been studied more than others. Scientists have had a considerable interest in Costa Rica and Panama, for example (Mabberley 1992). Soberingly though, fewer than half a million of the estimated several million organisms in the tropics have a Latin name, and a minute fraction are understood with respect to their ecology or even distribution. For at least 90% of all tropical species the ecological future is either extinction or restriction to reserves totalling less than 10% of the tropical land surface. Mabberley (1992) sums the situation up by saying "It is the duty of biologists to draw attention to the possible fate of what is left . . ."

THE CASE OF BOLIVIA

Of the biologically rich areas in the tropics, one country which has received very little attention scientifically either by inside or outside interests, is Bolivia. This chapter presents Bolivia as a case study in addressing the challenges of communicating ecology within a nation of extremely high biodiversity, faced with potentially unfettered resource exploitation in a climate of poor knowledge and education. In addressing ecological issues, Bolivia lacks biologists and few scientists are present with the knowledge to constructively contribute to

[1]School of Biological Sciences, Victoria University, P.O. Box 600, Wellington, New Zealand.
Present address: Department of Botany, University of Otago, P.O. Box 56, Dunedin, New Zealand.

Pp. 65–72 *in* ECOLOGY FOR EVERYONE: COMMUNICATING ECOLOGY TO SCIENTISTS, THE PUBLIC AND THE POLITICIANS ed by R. T. Wills and R. J. Hobbs. Surrey Beatty & Sons, 1998.

management decisions or develop conservation or sustainable management practices. Furthermore, the country has been largely neglected by the international scientific community almost as an accident of history, despite having some of the last large tracts of undisturbed lowland habitat left in South America.

Bolivia extends from latitude 10°S to 23°S and covers 1 098 581 km². It has a great variety of landscapes, stretching from the high Andes at around 6 500 m through the central valleys to the eastern lowlands where the city of Santa Cruz is located at about 450 m elevation (Fig. 1). In parts of the Department (= Province) of Pando in the north, the altitude is less than 120 m above sea level. About two-thirds of Bolivia lies in the lowlands of the Amazon Basin, but the majority of its 6.5 million people are concentrated in the western high altitude region (the altiplano — Fig. 1; Klein 1982) with much of the lowlands remaining untouched until recently. Because of land degradation, resource depletion and population increases, the people of the high Andes have started migrating downhill bringing with them cattle and a steady agricultural advance. Some estimate that Bolivia could "accommodate" as many as 35 million people with no strain on resources (Carter 1971). However, such increases would engender considerable losses in biodiversity, land degradation and resource depletion without a managed programme of conservation and sustainable management.

Consequently, efforts to communicate ecological knowledge and sustainable practices are critical particularly as the pressures to clear land are increasing exponentially both from inside and outside the country. Bolivia has a tumultuous political history with many short-lived governments. However, the National Revolution of 1952 was significant in that it resulted in social, economic and political reforms. At this time, the dominantly native Indian population were finally given political power, as well as their lands, allowing the country to evolve "in a manner distinct from the rest of the hemisphere" (Klein 1982). The nonassimilation of the native Indians following past colonizations by Inca then Spanish cultures, has given the country its strong linguistic and cultural identity (Malloy and Thorn 1971) and renders it distinct among the Latin American nations. Indeed, the human diversity is very great with the first language not Spanish but one of the many native Indian languages. Some estimate these to number 300, within around 20 major linguistic types (Dr S. Halloy, pers. comm. 1996). In many respects, the transfer of power after 1952 has contributed to the relatively low level of resource exploitation up until now in the lowland portions of Bolivia. Today, the country is perceived as a place with vast natural resources but the native Indian population remains extremely poor (Alexander 1982).

In the 1980s, Latin America had an estimated 673 715 × 10³ ha. of tropical closed forest of which Bolivia contained 44 010 × 10³ ha (approximately 15%) (Reid 1992). Annual rates of clearance in 1981–85 in Bolivia were

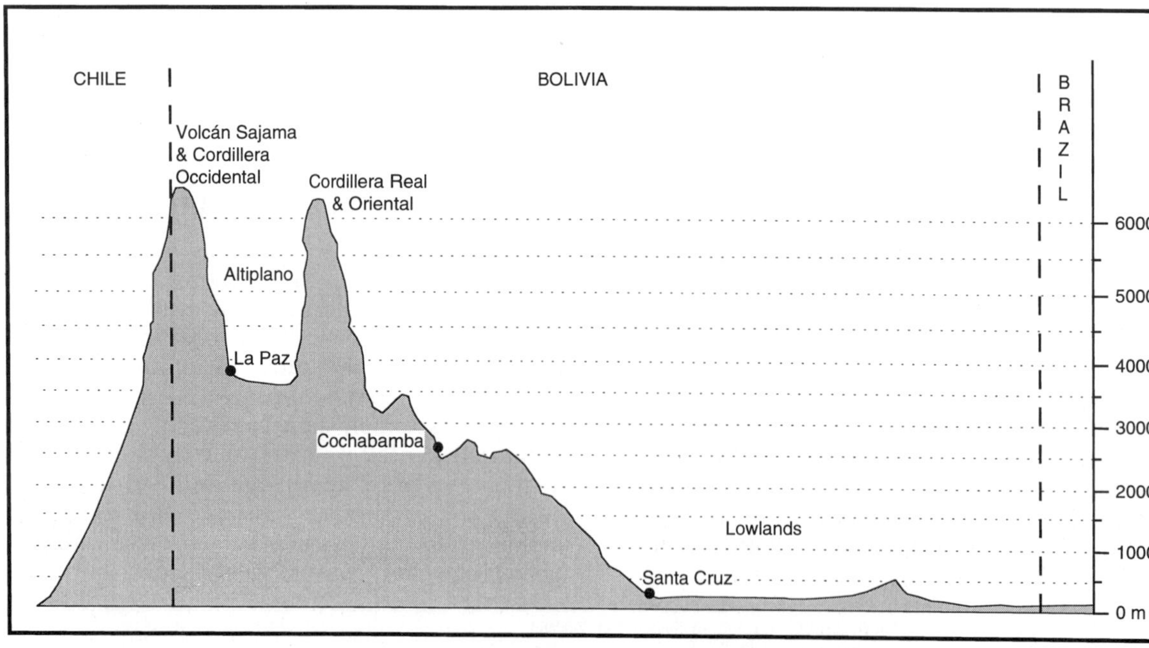

Fig. 1. A stylized cross-section through Bolivia from the coast of Chile in the west to Brazil in the east. Bolivia extends approximately 1 500 km from west to east.

estimated at 0.2% per year compared with 0.4% in Brazil, 1.8% in Colombia and 2.4% in Ecuador (Whitmore and Sayer 1992). Relatively then, the lowland Amazon rainforests of Bolivia are still unmodified. However, rates of clearance in general, have accelerated in recent years with smoke from clearance of Brazilian rainforests in 1995 being the worst on record according to NASA (Dr S. Halloy, pers. comm. 1995). Many of these fires spread far into Bolivian territory (author, pers. obs. 1995). This type of reduction in forest area forms the basis for current estimates of global species loss of up to 50% by the year 2000 (Ehrlich 1981 *cited in* Pearce and Moran 1994) representing a rate between 1 000 to 10 000 times the historical rate of extinction (Wilson 1988).

Bolivia presents one example of where individuals may not appreciate that their actions, when combined spatially and temporally, have a vast impact on the survival of flora and fauna. When population pressures increase, short-term gain may have more immediate importance than thought for longer-term perspectives. The most extensive and biologically destructive use to which these natural habitats are converted, is agriculture. Such activity can range from cropping to shifting cultivation, pasture conversion to wood extraction (Huston 1994). There are many examples of where such natural ecosystem destruction has not resulted in improved quality of life but a cycle of continuing poverty (Fearnside 1990). Huston (1994) ventures to suggest that "there is no inherent conflict between the preservation of biological diversity and the economic improvement of the human condition." His reasoning stems from the view that land with high species diversity is not a good candidate for profitable agriculture or even productive forestry (Huston 1994) despite many attempts to develop such land in this way, for example in the Brazilian Amazon Basin (Fearnside 1990). Without urgent and considered input, Bolivia could suffer such a fate with the apparently lush rainforest environment yielding little economic return once gone. In these circumstances, the importance of effective communication between the Bolivian people, scientists and governments cannot be overemphasised.

Numerous studies exist regarding the range of economic benefits that can result from conservation (Pearce and Moran 1994) and research is increasing on the sustainable management of tropical forest systems (e.g., Uhl *et al.* 1997). It is these types of endeavours that Bolivia must move toward if the country is to safeguard its natural heritage and truly benefit the population (H. Justiniano, pers. comm. 1995). Debt-for-nature swaps, encouragement of ecotourism, and commercialisation of genetic resources are all options being explored through the support of the US-based organization *The Nature Conservancy*, and private sponsors. By 1992, 20 debt-for-nature agreements totalling US$110 million had been arranged in nine countries worldwide including Bolivia (Wilson 1992; Panayotou 1994). Such agreements may entail writing off some or all of a country's national debt in return for protection of particular areas or habitats for conservation.

The role of non-governmental organizations

Bolivia is an example of where there is an urgent need for rapid education and effective communication of biological values and sound ecological practices. Given the lack of knowledge within the country outside aid has been sought. At present, one of the organizations with the greatest support from outside funding agencies and individuals is the Fundación Amigos de la Naturaleza (F.A.N.) which was founded in 1989 with the support of *The Nature Conservancy* with the primary objective being the conservation of biodiversity in Bolivia. F.A.N. currently has a mainly Bolivian staff of around 60 and works in three large protected areas including two National Parks (*Parque Nacional Amboró* and *P. N. Noel Kempff Mercado*) in the province of the Department of Santa Cruz (Fig. 2). The organization has specific management responsibilities, a role which involves guarding the reserved areas from illegal activities such as timber extraction, drug trafficking, hunting, agricultural encroachment and fishing. In effecting this management, F.A.N. aims to communicate with and involve local communities in guarding the parks. Further, it is hoped that the development of ecotourism will provide an income base thereby avoiding the lure of involvement with illegal profiteering.

As part of creating better management frameworks, a Science Department was established in F.A.N. in 1995. This department has taken on the task of establishing a scientific base for setting priorities for conservation activities in the existing protected natural areas, in cooperation with international scientists. The Science Department also aims to identify and establish further areas for protection in what constitutes a race against time. The Section is acutely aware that research priorities have to be established in order to correctly and effectively allocate funds. However, at present, almost any research proposed is allowed because of the need for as much information as possible to be gathered (A. Moreno, pers. comm. 1996).

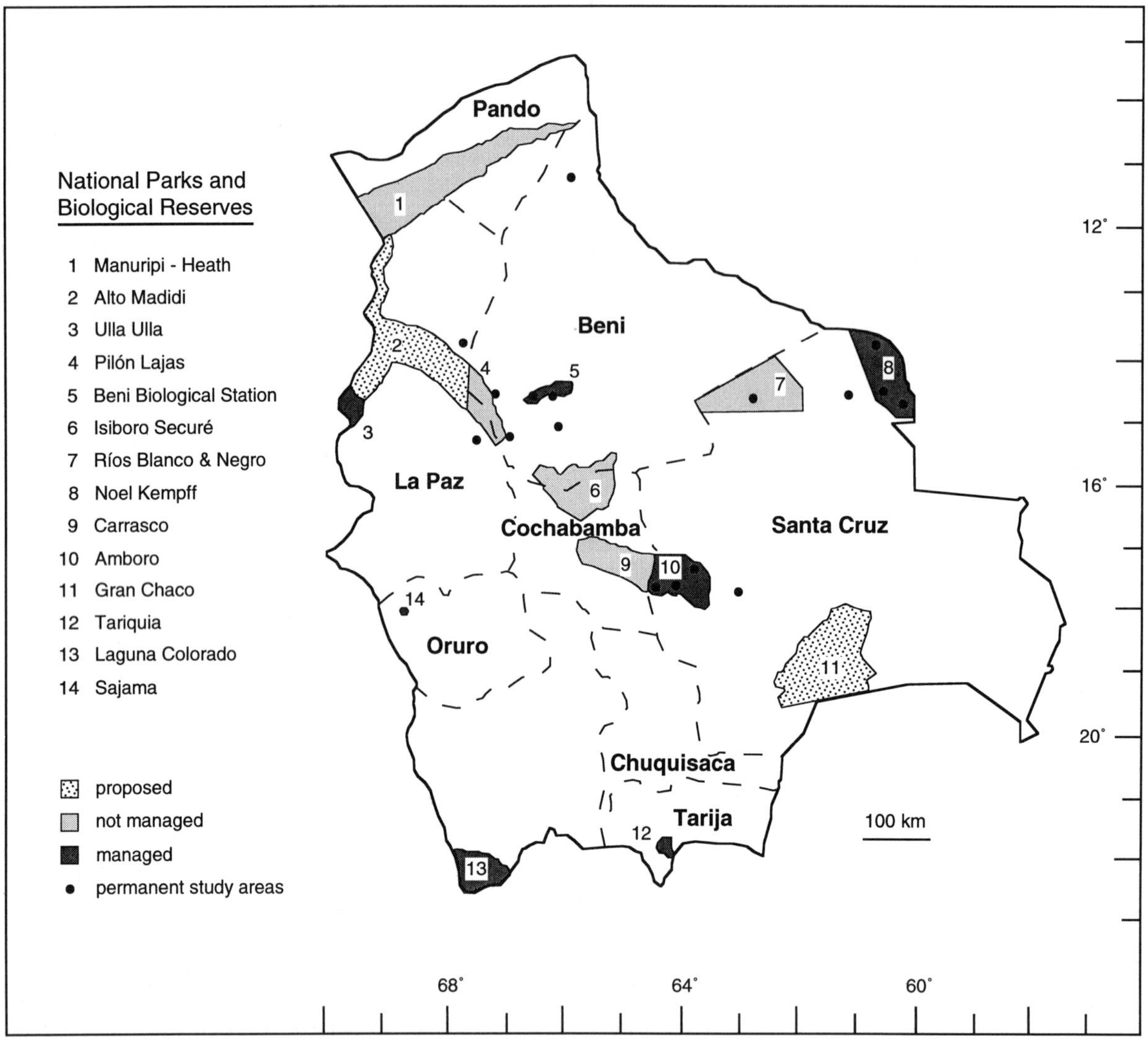

Fig. 2. The National Parks and Biological Reserves of Bolivia and their management status in 1995. The eight political provinces (Departments) are shown along with permanent study sites associated with the Museo de Historia Natural "Noel Kempff Mercado". *Parque Nacional Noel Kempff Mercado* (No. 8) and *P. N. Amboró* (No. 10) are both in the Department of Santa Cruz (adapted from Killeen unpub.).

Parque Nacional Noel Kempff Mercado

This Park illustrates the challenges, problems and realities involved with conservation and sustainable management in Bolivia. Until 1980, one of the least known regions of the country was the northeast section of the Department of Santa Cruz bordering with Brazil (13°30'–15°S and 60°30'–61°30'W). In 1979, 541 000 hectares of this area which forms part of the Precambrian Brazilian Shield, was declared the *Parque Nacional Huanchaca*. The Huanchaca Plateau which reaches an altitude of 900 m, is probably better known worldwide through the writings of Conan Doyle for it is the Lost World discovered by the explorer Colonel Percy Fawcett in 1910. In 1986, the Park became known as the *P. N. Noel Kempff Mercado* in memory of a prominent Bolivian scientist and conservationist who was killed by drug traffickers whilst doing fieldwork there. At this time, it was expanded to 706 000 hectares. Further expansion of the Park is currently under negotiation; if successful it will cover a total of 1. 6 million hectares. It is also adjacent to a few million hectares of forest reserve to the west, and a one million hectare cattle ranching area to the south.

When the *P. N. Noel Kempff Mercado* was created, little was known of its flora and fauna. Due to its position, the Park shares influences of the Amazon to the north, the Chaco to the south, the Brazilian Cerrado to the east and the Andes to the west. In these tropical latitudes (13°–15°S), this confluence presents a very high diversity of habitats and species. In addition, small changes in the topography and edaphic features can result in large differences in structure and composition of the vegetation, often reflecting past disturbance events such as fire and flooding. The plateau itself is dominated by Cerrado savannahs with patches of tall semi-evergreen gallery forest associated with creeklines (Fig. 3). Downslope, tall evergreen forest dominates with liana forests on the interfluves. The rivers in the region have extensive flood plains which are seasonally inundated supporting both forest and savannah.

Fig. 3. Tall lowland rainforest on the slopes of the Huanchaca escarpment. The large tree is mahogany (*Swietenia macrophylla:* Meliaceae), a highly prized timber species now extremely rare in the wild.

Knowledge and understanding of the Park's biological diversity is critical to its successful management for conservation. The almost completed Management Plan records 620 bird species from the Park, and this figure is likely to climb to 700 as further surveys are done. A staggering 25% of the bird species in the neotropics[1] are represented here. The mammalian fauna is abundant with many of the species which are rare or threatened within South America, for example giant otters, tapirs and pampas deer. Eighty-one species of non-flying mammals have been recorded, ranking as the highest number of these animals of any neotropical conservation area with available data. As to the floristic diversity, Bolivia as a whole is the least collected of all the South American countries although efforts are being made to redress this situation. The best estimate is around 18 000 vascular plant species — a figure likely to prove highly conservative. Given the range of habitats contained within *P. N. Noel Kempff Mercado,* it is reasonable to expect that the Park harbours a large floristic diversity. Four thousand vascular plant species have so far been recorded some of which are now rare through exploitation elsewhere, such as mahogany *Swietenia macrophylla* (Meliaceae: Fig. 4). The distribution of many taxa remains unknown though through the focused efforts of botanists affiliated with the Missouri Botanical Garden, a comprehensive taxonomic inventory is being developed.

In the north of the Park, a former estancia (cattle station) *Flor de Oro* (Flower of Gold) is the main base being developed for ecotourism. It sits alongside the Rio Guaporé, the river which forms the border with Brazil in this part of Bolivia. Biogeographically, this area is very

Fig. 4. The Huanchaca Plateau at around 800 m above sea level. Creeklines support fingers of diverse semi-deciduous gallery forest whilst the savannah has a mix of Cerrado (shrubland/low trees) and grassland habitat. A thunderstorm brews on the escarpment in the background.

[1] The neotropical region includes part of southern North America, Central and South America but excludes the *Nothofagus* forests of southwestern Chile which are considered to be in the Antarctic region.

Fig. 5. A matamata turtle *Chelus fimbriatus*, one of the very distinctive freshwater organisms which inhabit the inland river systems of north-east Bolivia. The turtle, considered rare, has been caught and has then died in an illegal fishing net confiscated by park guards, one of whom looks on. The animal is being injected with formalin to prepare for transport to the Museo de Historia Natural "Noel Kempff Mercado" in Santa Cruz. Collected in November 1995 from the Rio Guaporé, the river forming the border between Brazil and Bolivia.

distinct. The blackwater rivers for example, harbour freshwater dolphins, stingrays, sponges, turtles (Fig. 5), giant otters, alligators and piranhas whilst the forest and pampas are home to jaguar, monkeys, tapir and armadillos to name just a few. Many species found associated with the rivers are from groups normally associated with the marine environment which raises some interesting questions over the origin of the Amazon Basin. The setting is a very appealing one with ready access to river, wetland, forest and pampas grassland habitats. However, there are many management questions and difficulties to address, not least education of Park employees over exploitation of rare plants and introduction of exotic animals, as well as fundamental ecological questions such as the use of fire as a conservation management tool.

An integrated research programme is currently being proposed for *P. N. Noel Kempff Mercado* (Dr T. Killeen, pers. comm. 1996). Studies are focusing on the forest-savannah boundary, in a similar way to a number of northern Australian studies (*c.f.* Bowman 1992, 1993; Bowman and Fensham 1991) with the long-term goal to monitor the impact of climate change on biological diversity in this climatic transition zone. An integral part of this research and management programme is the development of comprehensive vegetation maps (*c.f.* Wilson *et al.* 1990).

Options and challenges

Protected areas such as *P. N. Noel Kempff Mercado* and others in the Department of Santa Cruz do not fully represent the diversity of habitats within the Department or Bolivia as a whole. It is essential that research, survey and monitoring programmes are established for other protected areas similar to that being developed for *P. N. Noel Kempff Mercado*. In doing so, an integrated management approach is essential between individuals, institutions and agencies both within and between the Departments.

The management challenges are many. Protected Area status does not necessarily guarantee protection and in some instances reserved areas have been revoked as in the case of *P. N. Amboró* (Figs 2, 6). This Park is suffering the more immediate effects of human migration which is reflected in its recent political history. The original 1984 dedication of 180 000 ha was increased to 637 600 ha in 1991 but then decreased to 420 000 ha in October, 1995. The reduction reflects the acute conflict between conservation and development. Poverty combined with a lack of understanding and education contributes to the general population being unable to appreciate the negative effects of widespread cattle and horses, repeated use of fire and land clearance. As a consequence, the designated buffer zones of the Park have presented a major administrative and policing problem. F.A.N. is not in charge of this Park at the moment but are committed to actively participating in its management in order to guarantee its conservation (A. Moreno, pers. comm. 1996). Some other protected areas in the country also have little status apart from on paper (Fig. 2). *P. N. Noel Kempff Mercado* in some respects has been protected by its isolation in Bolivia but the dangers of incursion from Brazil pose a serious problem (Dickinson 1995[1]).

Biodiversity options and sustainable management practices in Bolivia are unlikely to be effective unless the native Indian population is

[1]Dickinson, K. J. M., 1995. *Ecological and Management Issues in Protected Natural Areas under F.A.N.'s Jurisdiction*. Unpublished Report to F.A.N., December 1995, Santa Cruz, Bolivia.

Fig. 6. The landscape of the *Parque Nacional Amboró*. At around 3 000 m above sea level, these mountains are vegetated by floristically diverse cloud forest. This Park is under severe pressure from population increase and incursion. Cattle and their impact can be found in a substantial portion of the Park.

part of the solution and not excluded from it. In practice, this means involving people in the identification of their priorities and needs, as well as in project design, implementation and evaluation (Wells 1994). One of the biggest challenges then, is the need to communicate the national and international significance of Bolivia's natural values to its general population. There will always be the potential for conflict of interest between the peoples' ability to earn a living or subsist and the conservation of areas of high ecological value. Conservation and sustainable management projects can aim to mitigate such conflicts by effectively promoting and communicating ways of generating alternative sources of income as well as through education programmes.

There are now many examples of projects endeavouring to use local economic incentives for the protection or sustainable use of biodiversity (see Wells 1994). In Bolivia's case, high hopes are held for the development of ecotourism as a creditable means for boosting the local economy and providing new funds for conservation. Raising some of the protected areas to an international category of reserve status would also be likely to benefit Bolivia's profile. For example, World Heritage listing or Biosphere Reserve status could be sought for the country's top class Parks with their very rich biological and landscape diversity. A raised profile would assist marketing of ecotourism more widely though the additional challenge is to manage the costs and benefits associated with such activity. There is the pitfall of tourism resulting in excessive damage of the areas being protected (*c.f.* Wells 1993). Whether ecotourism can provide the bonanza of funding required to safeguard the natural resources of Bolivia long enough to allow the development of comprehensive management plans remains to be seen. Other income options for the country currently being investigated are biological prospecting and sustainable use of biological products though as Wells (1994) points out, limits for exploitation of the latter are not only difficult to determine but also difficult to regulate.

For this biologically rich region, the future presents major challenges. Education involving the communication of the global significance of this country's biodiversity and the development of sustainable management practices takes time, commitment and resources. There is a clear role for the international scientific community to work with Bolivian scientists and managers, in the development of research strategies and management programmes for biological conservation and sustainable management. Undoubtedly, the key to the success of these programmes is effective communication at both a national and international level.

ACKNOWLEDGEMENTS

I thank Victoria University on behalf of F.A.N. and myself for granting me the special leave enabling me to accept the invitation to work in Bolivia. I appreciate the foresight of F.A.N.s Executive Director, Hermes Justiniano, in encouraging international scientific involvement in the pressing ecological and conservation issues facing Bolivia. F.A.N. provided logistical support for me whilst in the country. Dr Tim Killeen of the Museo de Historia Natural "Noel Kempff Mercado" and the Missouri Botanical Garden invited me to work in the *P. N. Noel Kempff Mercado.* Dr Stephan Halloy and Maria Lidia Hance deserve special thanks for their considerable support during my time in the country. Dr Stephan Halloy, Dr Alan Mark, Dr Anya Waite and two anonymous referees also provided helpful comments on the manuscript.

REFERENCES

Alexander, R. J., 1982. Bolivia. Past, Present and Future of Its Politics. Hoover Institution Press, Stanford University, California.

Bowman, D. M. J. S., 1992. Monsoon forests in north-western Australia. II. Forest-savanna transitions. *Aust. J. Bot.* **40:** 89–102.

Bowman, D. M. J. S., 1993. Establishment of two dry monsoon forest tree species on a fire-protected monsoon forest-savanna boundary, Cobourg Peninsula, northern Australia. *Aust. J. Ecol.* **18:** 235–37.

Bowman, D. M. J. S. and Fensham, R. J., 1991. Response of a monsoon forest-savanna boundary to fire protection, Weipa, northern Australia. *Aust. J. Ecol.* **16:** 111–18.

Carter, W., 1971. Bolivia. A Profile. Praeger Publishers, New York.

Daily, G. C. and Ehrlich, P. R., 1994. Population extinction and the biodiversity crisis. Pp. 43–53 *in* Biodiversity Conservation ed by C. A. Perrings, K. G. Mäler, C. Folke, C. S. Holling and B. O. Jansson. Kluwer Academic Publishers, The Netherlands.

Dobson, A. P., 1996. Conservation and Biodiversity. Scientific American Library, New York.

Ehrlich, P. R. and Ehrlich, A. H., 1990. The Population Explosion. Simon and Schuster, Sydney.

Fearnside, P. M., 1990. Deforestation in Brazilian Amazonia. Pp. 211–38 *in* The Earth in Transition: Patterns and Processes of Biotic Impoverishment ed by G. M. Woodwell. Cambridge University Press.

Huston, M. A., 1994. Biological Diversity. The Co-existence of Species on Changing Landscapes. Cambridge University Press.

Hutching, G. and Potton, C., 1987. Forests, Fiords and Glaciers: New Zealand's World Heritage. The case for a South-West New Zealand World Heritage Site. Royal Forest and Bird Protection Society, Wellington, New Zealand.

Klein, H. S., 1982. Bolivia. The Evolution of a Multi-Ethnic Society. Oxford University Press.

Mabberley, D. J., 1992. Tropical Rain Forest Ecosystems. Chapman and Hall, New York.

Malloy, J. M. and Thorn, R. S., 1971. Beyond the Revolution. Bolivia since 1952. University of Pittsburgh Press.

Panayotou, T., 1994. Conservation of biodiversity and economic development: the concept of transferable development rights. Pp. 288–305 *in* Biodiversity Conservation ed by C. A. Perrings, K. G. Mäler, C. Folke, C. S. Holling and B. O. Jansson. Kluwer Academic Publishers, The Netherlands.

Pearce, D. and Moran, D., 1994. The Economic Value of Biodiversity. IUCN with Earthscan Publications Ltd, London.

Peat, N., 1994. Manapouri Saved: New Zealand's First Great Conservation Success Story. Longacre Press, Dunedin, New Zealand.

Perrings, C. A., Mäler, K. G., Folke, C., Holling, C. S. and Jansson, B. O., 1994. Unresolved issues. Pp. 321–28 *in* Biodiversity Conservation ed by C. A. Perrings, K. G. Mäler, C. Folke, C. S. Holling and B. O. Jansson. Kluwer Academic Publishers, The Netherlands.

Reid, W. V., 1992. How many species will there be? Pp. 55–73 *in* Tropical Deforestation and Species Extinction ed by T. C. Whitmore and J. A. Sayer. Chapman and Hall, London.

Shiva, V., 1988. Staying Alive: Women, Ecology and Development. Zed Press, London.

Smith, S. J. and Banks, M. R. (eds), 1993. Tasmanian Wilderness — World Heritage Values. Royal Society of Tasmania, Hobart.

Uhl, C., Barreto, P., Veríssimo, A., Vidal, E., Amaral, P., Barros, A. C., Souza, Jr C., Johns, J. and Gerwing, J., 1997. Natural resource management in the Brazilian Amazon. *Bioscience* **47:** 160–68.

Wells, M. P., 1993. Neglect of biological riches: the economics of nature tourism in Nepal. *Biol. Cons.* **2:** 445–64.

Wells, M. P., 1994. Biodiversity conservation and local development aspirations: new priorities for the 1990s. Pp. 306–20 *in* Biodiversity Conservation ed by C. A. Perrings, K. G. Mäler, C. Folke, C. S. Holling and B. O. Jansson. Kluwer Academic Publishers, The Netherlands.

Whitmore, T. C. and Sayer, J. A., 1992. Deforestation and species extinction in tropical moist forests. Pp. 1–14 *in* Tropical Deforestation and Species Extinction ed by T. C. Whitmore and J. A. Sayer. Chapman and Hall, London.

Wilson, B. A., Brocklehurst, P. S., Clark, M. J. and Dickinson, K. J. M., 1990. Vegetation Survey of the Northern Territory, Australia. Conservation Commission of the Northern Territory, Technical Report, No. 49, Darwin.

Wilson, E. O., 1988. The current state of biological diversity. Pp. 3–18 *in* Biodiversity ed by E. O. Wilson. National Academy Press, Washington.

Wilson, E. O., 1992. The Diversity of Life. Harvard University Press, Cambridge, Massachusetts.

World Commission on Environment and Development, 1987. Our Common Future. Oxford University Press.

CHAPTER SIXTEEN

Algorithms, politics and timber: an example of the role of science in a public, political negotiation process over new conservation areas in production forests

R. L. PRESSEY[1]

INTRODUCTION

THE conservation of biodiversity relies almost entirely on *in situ* protection — the ability of one species, *Homo sapiens*, to manage parts of the landscape and marine environment in a way that separates populations, species, assemblages and ecosystems from processes that threaten their persistence in the wild. The science of conservation planning aims to put the right conservation measures, as far as possible, in the right places. A conservation measure is right if it provides enough protection from threatening processes to maintain the natural features of concern. A place is right if it has priority for conservation action to promote the persistence of features that would otherwise decline or disappear.

Inevitably, conservation planning involves more than science. Decisions on the locations of new conservation areas are often highly political. The costs of the necessary measures can be very high in terms of money, opportunities for other land uses, and social change. Some members of the voting public and lobby groups will be pleased with an outcome, others will be loudly displeased, and the distribution of those in each category in relation to electoral boundaries can determine the vigour with which governments pursue conservation initiatives. The best political solution can therefore be bad scientifically and vice versa. A reality of conservation planning is an ongoing tension between science and politics. Science is no substitute for politics as the art of the possible but, if conservation planning were solely political, the most expedient decisions would almost always be made, with very serious implications for the future of biodiversity (Pressey 1994; Pressey and Tully 1994). The science of conservation planning has also tended to exclude the public, even those people directly affected by the outcomes. From the perspective of the layperson, the goals of nature conservation have tended to be obscure and the methods, especially in the computer age, unfathomable.

This chapter concerns a new approach to resolving conservation issues. It was used in eastern New South Wales in the first half of 1996 and was based on a very transparent, scientific method for conservation planning. The method was applied to a contentious environmental issue and involved a variety of interest groups, with the implementation of its results resolved politically. The issue was the future use of about 2.4 million hectares of production forests in eastern New South Wales. The debate over logging versus conservation of the eastern forests of the State has been long and bitter (Watson 1990; Taplin 1992; Bonyhady 1993; Dargavel 1995). The region's forests support one of the largest concentrations of species in Australia and much of the remaining forest is extensive, wild and charismatic. Moreover, the State forests used for timber production in this part of New South Wales, like conservation reserves, have generally been dedicated over land remaining after conversion to private tenure and clearing for grazing and cropping (Lunney and Leary 1988; Lunney and Moon 1988; Richards *et al.* 1990). Conflict between forestry and conservation interests was therefore inevitable.

The chapter begins by describing a scientific idea looking for a role in the real world. The idea is called irreplaceability, a measure of the importance of any area for achieving an explicit regional conservation goal. The chapter discusses the potential applications of irreplaceability in conservation planning, then describes the process of policy development at the Commonwealth and State levels that made

[1]New South Wales National Parks and Wildlife Service, P.O. Box 402, Armidale, New South Wales, Australia 2350.

Pp. 73–87 *in* ECOLOGY FOR EVERYONE: COMMUNICATING ECOLOGY TO SCIENTISTS, THE PUBLIC AND THE POLITICIANS ed by R. T. Wills and R. J. Hobbs. Surrey Beatty & Sons, 1998.

its application first possible, then necessary. This is followed by an account of a computer-assisted negotiation process over the extent and location of new conservation areas, a summary of the outcomes, and some comments on the future of this approach to regional planning.

THE CONCEPT OF IRREPLACEABILITY

In 1994, one of the outcomes of research on conservation planning was an idea called irreplaceability (Pressey *et al.* 1994). It was developed in response to the limitations of reserve selection algorithms used at the time. A reserve selection algorithm is a computer programme applied to the problem of finding the smallest possible set of areas that will achieve a nominated conservation goal. Examples of conservation goals are to represent (or contain) at least one occurrence of every species or at least 5% of the total area of every vegetation type in a region. The set of areas needed to represent at least one occurrence of each of the land systems in part of the Western Division of New South Wales is shown in Figure 1.

The type of result shown in Figure 1 can be useful as an indication for comparison with another conservation goal but, for real-world planning, it fails to convey important information. It leaves people without an understanding of the potential importance of the areas not selected by the algorithm and gives no indication of the possibilities of replacing some selected areas with others if they prove not to be available or feasible for nature conservation. This is a particular problem for planners and managers who might have to deal with landholders on the selected areas. All they know about any particular area is that an algorithm told them it was needed to achieve a conservation goal — they have no regional perspectives on relative importance. Such a result also lacks credibility with politicians and the general public. Data were fed into a computer, a series of analytical steps was carried out, and a result emerged. It is fair to ask what happened between the input and the output and whether the outcome might have been different if the selection process had been more closely overseen or managed by humans. Irreplaceability analysis is a solution to these

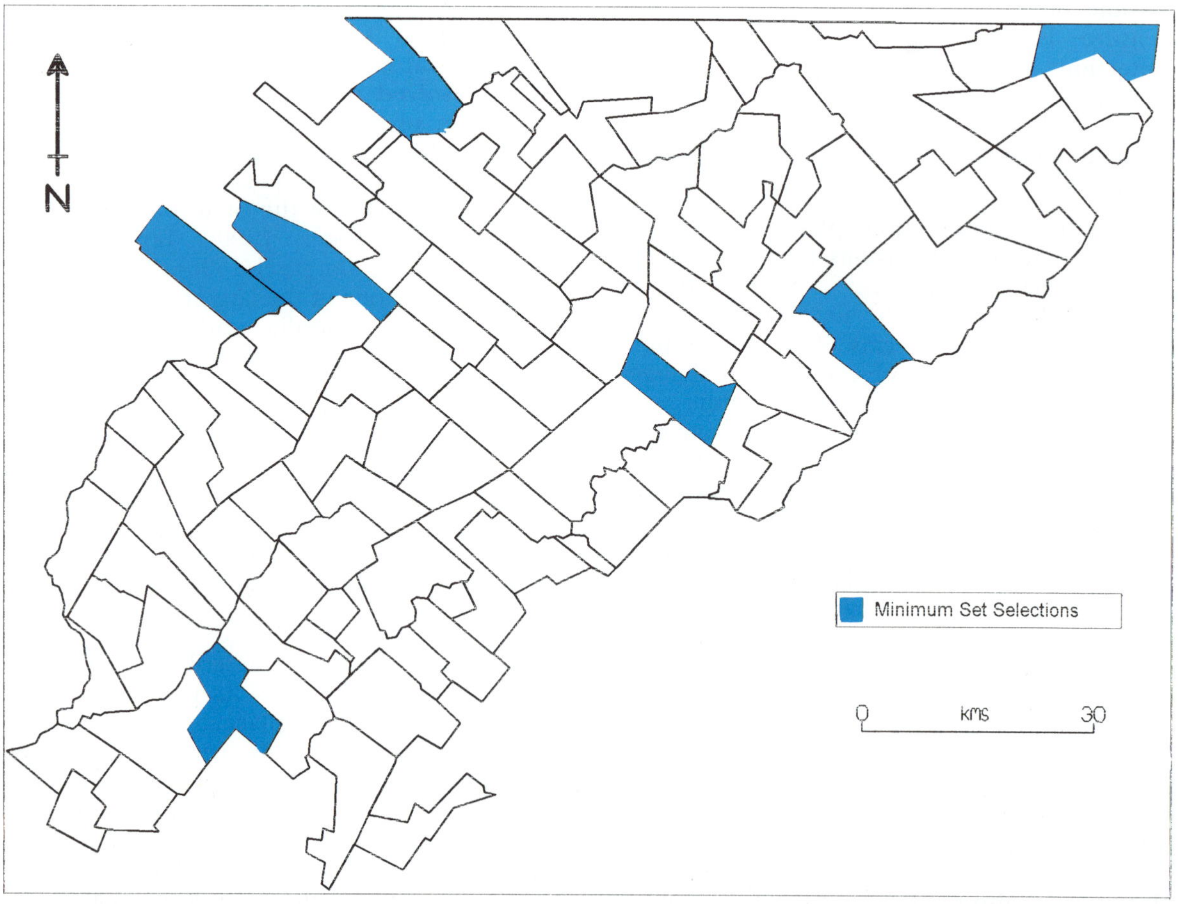

Fig. 1. An environmental region in the Western Division of New South Wales (Province 1Ac of Morgan and Terrey 1990) consisting of 95 pastoral holdings with a total area of 6 625 km^2. The northern edge is the Queensland border. The seven holdings coloured blue were selected by an algorithm to represent at least one occurrence of each of the land systems in the region. Modified from Pressey *et al.* (1993).

limitations. It gives a picture of the relative importance of every area in the region (Fig. 2) and allows planners, managers and others to explore the options for achieving a conservation goal.

Irreplaceability can be defined in two ways:

1. The likelihood that any of the areas in a region will be needed to achieve an explicit conservation goal. Some of the pastoral holdings in that region are completely irreplaceable (values of 100) (Fig. 2). No matter how a system of conservation areas is put together, the bright red areas will be needed as part of that system. They have no replacements. Other holdings are replaceable to varying degrees. Those with higher values have fewer replacements. Those with lower values have many replacements. This opens up options for constructing systems of conservation areas that achieve the goal of representing at least one occurrence of each land system in the region but which differ in configuration, land use history, acquisition cost and other factors.
2. The extent to which the options for achieving an explicit conservation goal are narrowed if any of the areas in a region is destroyed or made unavailable for conservation. An application for a clearing license on one of the totally irreplaceable holdings in Figure 2 has serious consequences. It would prevent the regional conservation goal from being achieved. Clearing of holdings with minimum irreplaceability, on the other hand, would have little effect on the options for achieving the goal. The gradient of irreplaceability values is therefore a guide for assessing the impacts on a regional conservation goal of clearing and other developments.

WAYS OF APPLYING IRREPLACEABILITY IN CONSERVATION PLANNING

The two definitions of irreplaceability are both ways of placing any particular area in a geographical context in terms of its importance for achieving a conservation goal. The context can be as wide or as narrow as planners and managers wish — a single local government area, a biogeographical region, the whole of New South Wales, or a whole continent. The irreplaceability of any part of a region depends not only on this context but very much on the

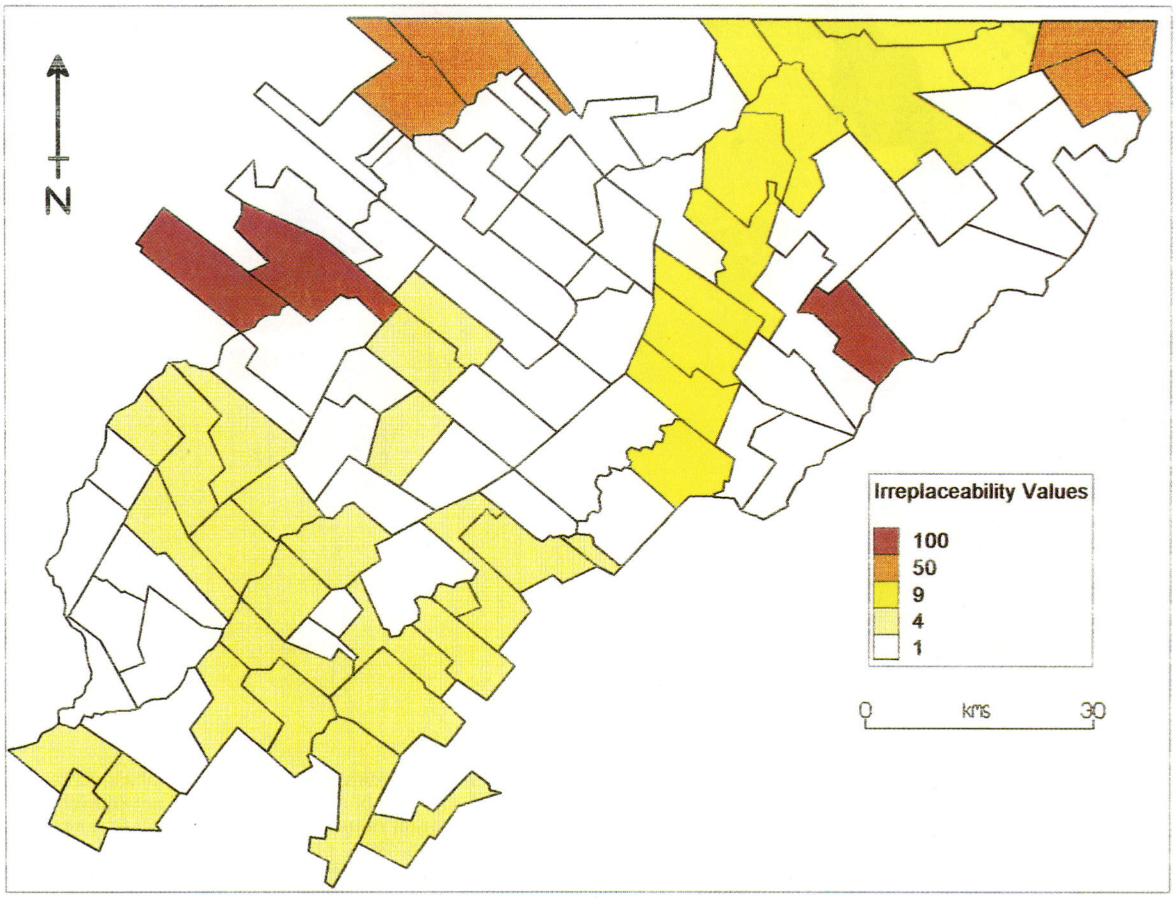

Fig. 2. The same environmental region as in Figure 1 showing the irreplaceability of all 95 pastoral holdings for the same conservation goal. Modified from Pressey *et al.* (1993).

conservation goal that has been set for the region. For example, the goal of reserving 75% of all the forest types, including their old growth phases, in part of northern New South Wales produces a pattern of irreplaceability of logging compartments (Fig. 3a). If the goal is changed to 85% of each of the same forest types, the pattern of irreplaceability changes (Fig. 3b). More compartments have higher levels of irreplaceability and there are fewer options in the region for constructing a set of conservation areas that will achieve the higher goal. The options are narrowed further if the goal is 85% of all of the same forest types and 85% of all the predicted faunal distributions (Fig. 3c). Irreplaceability analysis is therefore a way of illustrating the implications of alternative conservation goals.

Irreplaceability can, but does not have to, integrate the importance of a particular area across all the forest types, species or other features in the study region. It can also be used to identify the options for conserving individual features, for example, the option of conserving 85% of only forest type 46 (Fig. 3d). Irreplaceability patterns could also be produced for each of the other forest types and species of plants and animals in this region, but the results would be difficult to

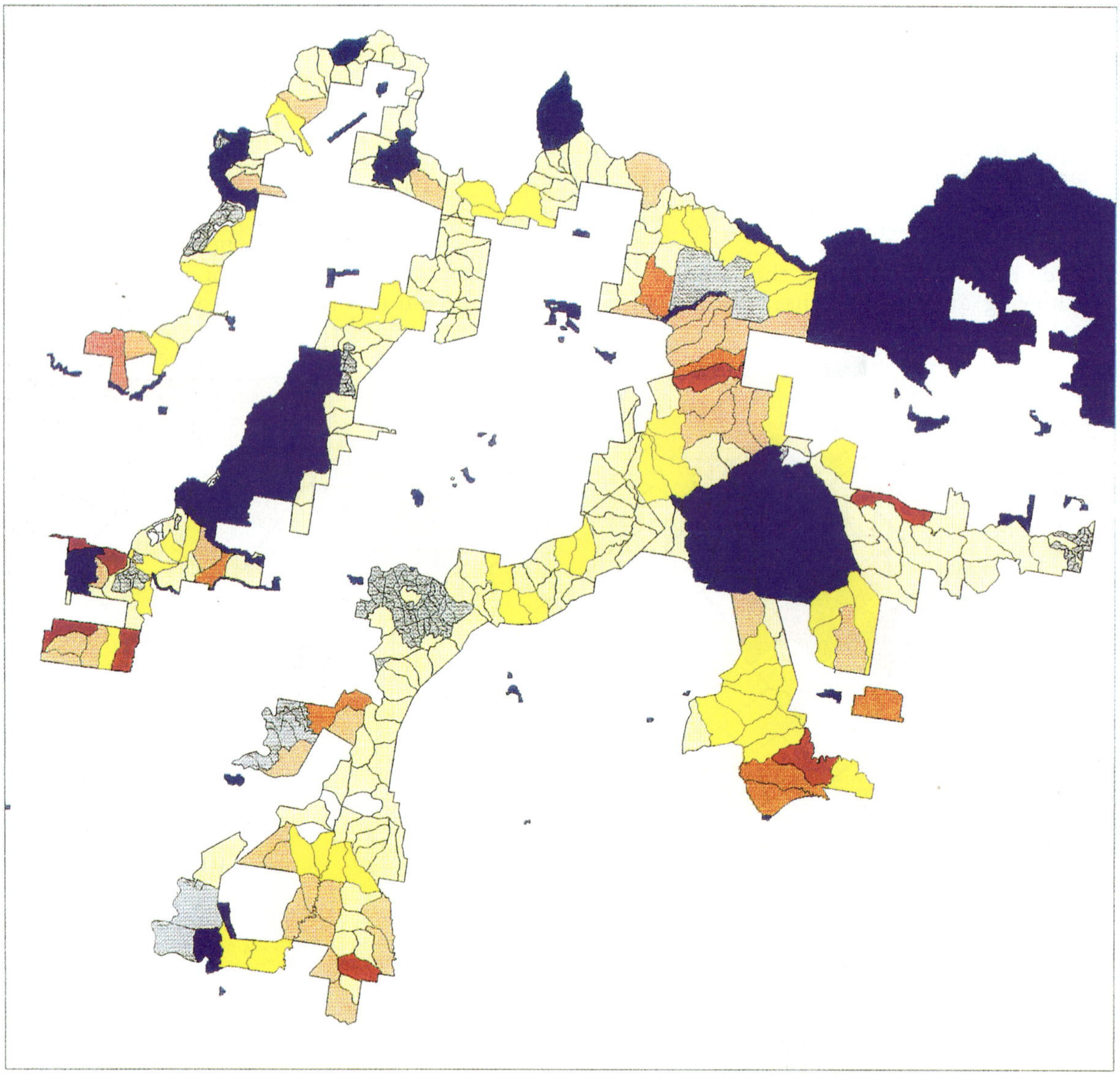

Fig. 3. Part of the N1 negotiation region for the Interim Assessment Process, showing irreplaceability values of logging compartments for four different conservation goals. Existing reserves are dark blue. Grey areas were not considered in the Interim Assessment Process either because of tenure or use for plantation timber. Irreplaceability values range from dark red (100% irreplaceability) through pink (80–99%), dark orange (60–79%), pale orange (40–59%), dark yellow (20–39%), pale yellow (>0–19%) and white (zero contribution to conservation goal). Patterns of irreplaceability correspond to the following conservation goals: (a) 75% of each forest type (this page); (b) 85% of each forest type (p. 77); (c) 85% of each forest type (p. 78) and 85% of the modelled distribution of selected faunal species; and (d) 85% of forest type 46 (p. 79).

integrate visually or manually. It is usually more efficient to allow the analysis to integrate features either across the whole data base or into groups such as forest types, fauna and flora.

If irreplaceability is to be used effectively in conservation planning, one more conceptual and analytical step has to be taken. As decisions are made on which of the optional areas will be given protection for conservation, the irreplaceability of some of the other areas in the region changes. This is because some of the features they contain have just been protected, either fully or partially. These unprotected areas therefore have less to offer to the regional conservation goal and are less irreplaceable. This calls for an interactive approach to using irreplaceability in real-world planning. Such an approach should lay out the options, accept decisions, recalculate irreplaceability values, and display the new set of options while showing clearly which areas have already been set aside for conservation. Ideally, the analysis would do all this quickly, preferably in seconds but certainly taking no more than a minute or two.

A working prototype of an interactive irreplaceability system was developed in May 1994 (Pressey *et al.* 1995). It was seen at the time as a basis for negotiating conservation plans with concerned parties gathered around a computer monitor to explore the options for achieving one or more explicit goals. As well as recalculating and redisplaying, it also helped the user to decide on the best options by providing any background information that was available on particular areas. Design decisions such as grouping of adjacent areas, buffering of existing reserves, and linking discrete conservation areas could be made after visual inspection of the juxtaposition of existing reserves and possible new conservation areas

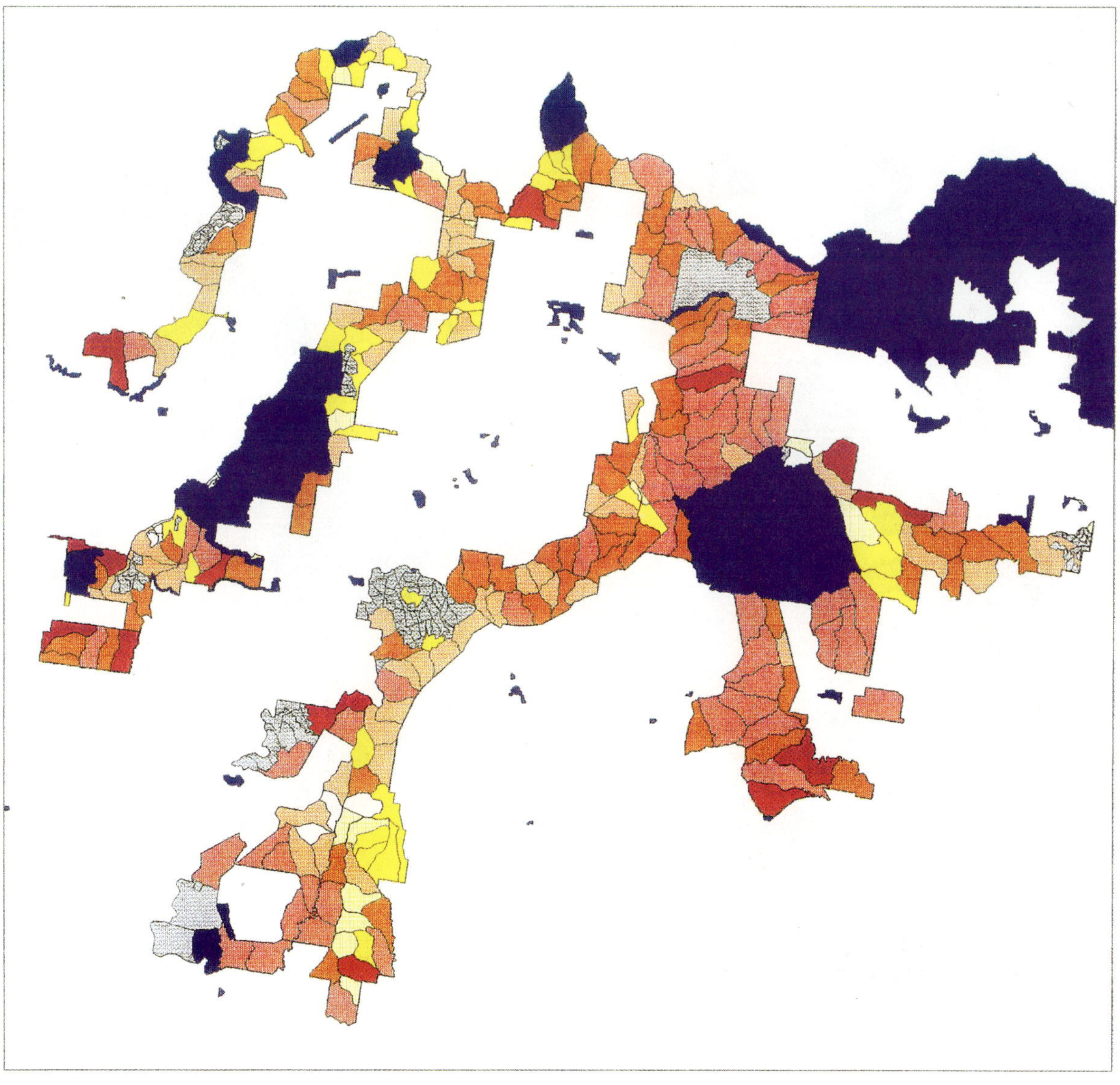

Fig. 3b. 85% of each forest type (see p. 76).

on the computer screen. The design of conservation areas is actually facilitated by information on irreplaceability because it identifies highly irreplaceable areas that are effectively fixed in the landscape and that will serve as nodes around which areas of lower irreplaceability can be grouped.

Good decisions on design also require good decisions on the size and boundaries of the individual areas being considered for conservation. In Figures 1 and 2 the areas are dictated by the boundaries of ownership and have an average extent of about 70 km^2. The logging compartments in Figure 3 average about 250 ha. The choice of size can be a trade-off between competing factors. Very large areas will make more viable and manageable reserves and can be easily amalgamated into even larger reserves, but will tend to represent natural features inefficiently by containing much larger areas of some forest types and species than are being targeted for conservation. Very small areas will represent natural features very efficiently, but will not be viable reserves in their own right. They will also be difficult to amalgamate into viable reserves during a planning exercise because of their sheer number and the patchy distribution of small areas with high irreplaceability.

By May, 1994, the concept of irreplaceability had been well developed, an interactive system to apply it in real-world planning was ready to use, and a range of other research issues on design and efficiency of representation were understood. The results of research were waiting to be used. While all this was happening, a parallel process of policy development was underway that would make it first of all possible, then necessary, to apply the research to a major and contentious conservation issue.

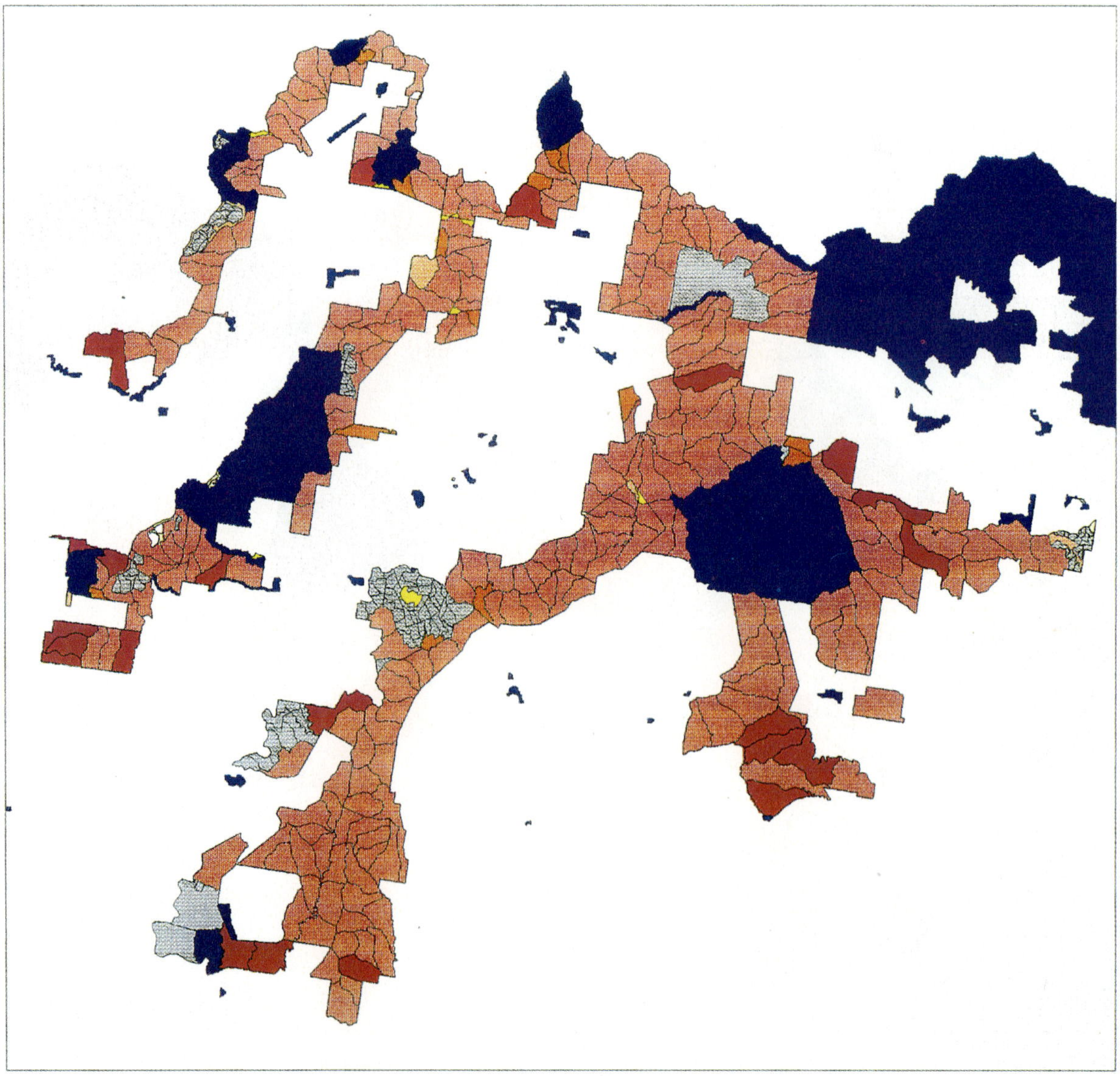

Fig. 3c. 85% of each forest type and 85% of the modelled distribution of selected faunal species (see p. 76).

THE POLICY FRAMEWORK THAT MADE IT POSSIBLE TO USE IRREPLACEABILITY IN REAL-WORLD PLANNING

One of the early lessons of research on systematic reserve selection is that the development of what seems a promising and important idea does not cause the world to beat a path to one's door. There are various reasons for this, mainly to do with the conditions that encourage *ad hoc,* rather than strategic, decisions on the allocation of conservation resources (Pressey 1995) and the lack of effective criteria for judging the real adequacy of a reserve system and for holding its designers accountable. The outcome is that, if researchers do not actively influence the policy of conservation planning, their research does not matter.

In April and May, 1996, irreplaceability analysis was central to the negotiations over new conservation areas in the production forests of eastern New South Wales. It was important in the process not because politicians and senior bureaucrats read the scientific publications arising from its development but because strategic policy on forest conservation, drawing on available knowledge in conservation biology and conservation planning, had been progressively refined since 1992. For brevity, this process of policy development is summarized as four steps, below.

1. The National Forest Policy Statement (Anon. 1992) established the principles of comprehensiveness, adequacy and representativeness as a basis for the location and design of forest reserves. Comprehensiveness concerned the need for the full variety of forest types or environments to be included in a system of reserves. Adequacy dealt with issues such as how much of each forest type should be reserved, in what configuration,

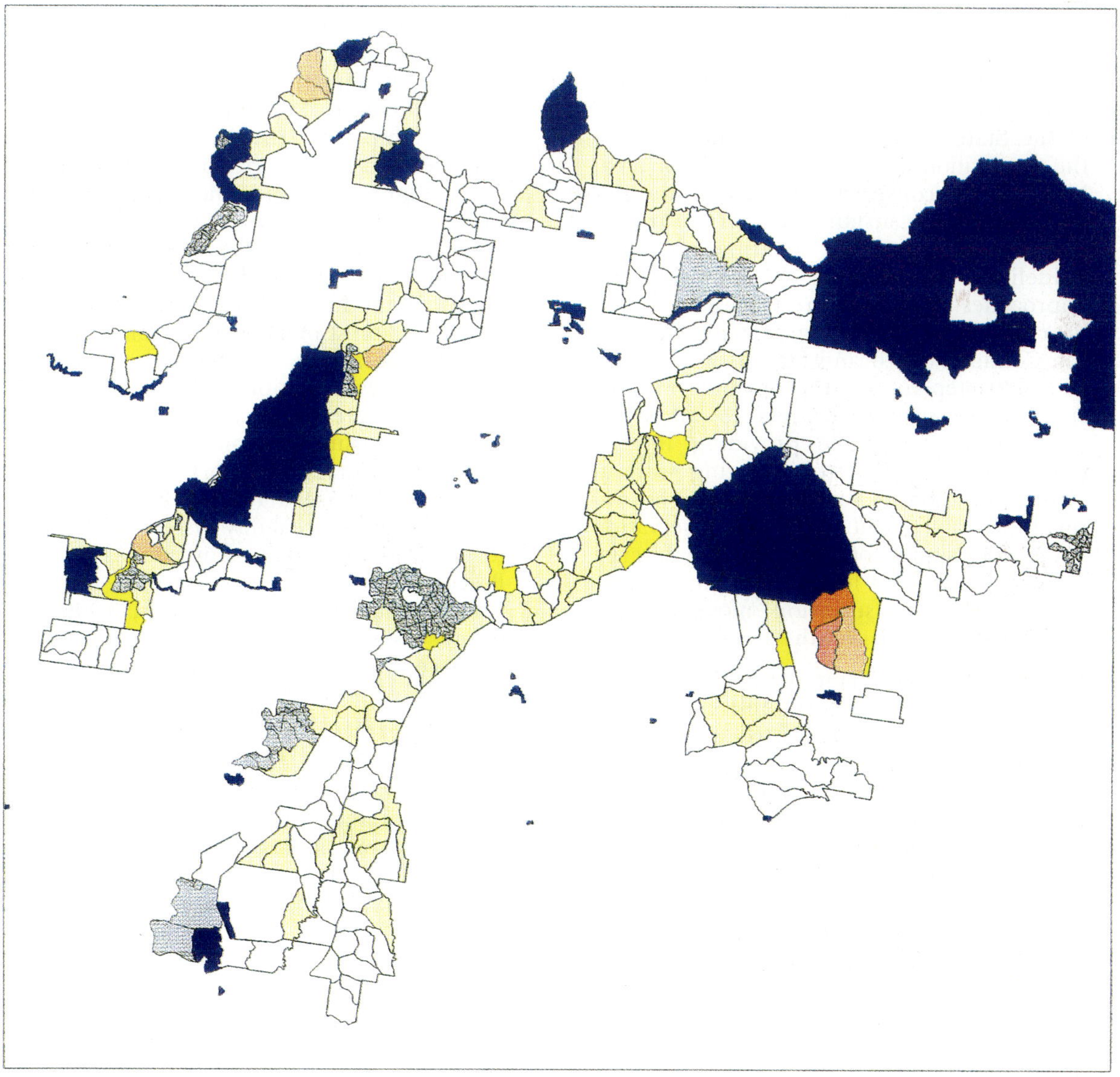

Fig. 3d. 85% of forest type 46 (See p. 76).

and with what pattern of replication. Representativeness referred to the importance of reserving the range of biotic diversity within forest types. These principles were endorsed by the Commonwealth, State and Territory Governments. At that time, the principles were far too broad to be made operational for conservation planning. Their importance lay in their elevation of the debate over the reservation of Australian forests above the traditional arguments over how many more reserves or how many more hectares were needed, statistics that can be meaningless in themselves. The principles concerned what should actually be in the new reserves.

2. In March, 1995, the Labor party came back into power in New South Wales. In June, 1995, the new government announced its forestry reform policy and outlined how it would meet its commitments to implementing the National Forest Policy Statement. The outcome of these commitments was to be a series of Comprehensive Regional Assessments that would identify the economic, social, cultural and natural values of the State's forests and develop plans for the reservation or other protection of areas having high conservation value while also maintaining a sustainable native timber industry. The resulting conservation plans, guidelines for forestry operations, and approaches to industry restructuring would be covered by formal agreements between the Commonwealth and State Governments. The first step toward the outcome was the Interim Assessment Process to allow for ongoing logging while providing interim protection (deferral from logging without change of tenure) for areas of high conservation value until the final conservation plans were signed off. Initially, the Interim Assessment Process focused on the familiar icons of the conservation movement — wilderness and old growth.

3. In July, 1995, the Commonwealth produced its proposed criteria for forest reservation (Anon. 1995) which attempted to put flesh on the framework of the broad principles laid out in 1992 in the National Forest Policy Statement. In interpreting the principles of 1992, the Commonwealth criteria dealt not only with wilderness and old growth but with biodiversity at the level of forest types and species. Some time later, the State Labor Government agreed to expand the scope of the Interim Assessment Process considerably to cover all biodiversity values. The interim process remained, however, focused only on public production forests and mostly east of the Great Dividing Range. The boundary of the assessment also enclosed the major and best known forest areas in the State and very large areas of old growth and wilderness. A political reason for dealing with the eastern regions was to concentrate on the forests from which woodchips are exported and which must be covered by export licenses from the Commonwealth. The focus on public land effectively delayed conservation assessments of private and leasehold forests in the east of the State which constitute up to half the remaining forests in some regions. The focus on the east put on hold assessments of forests, albeit limited in extent, in the drier western regions.

4. The development of the State Government's forestry reforms had involved extensive consultation with industry, unions and conservation groups. As the Interim Assessment Process gathered pace, this consultative process became even more important. A body called the Resource and Conservation Assessment Council was given the task by the State Government of co-ordinating the whole process of forestry reform, including the Comprehensive Regional Assessments. Its membership was drawn from the major agencies and interest groups with concerns in forest planning. It became clear that some form of negotiated planning was essential to resolve the conflicts between the different uses of forests and the various expectations about their future. Particularly when all aspects of forest biodiversity became involved in the assessment, it also became obvious that an analytical approach was needed that could integrate complex information into a picture of value for conservation that was easy to understand and work with. A decision was made to use a refined version of the prototype irreplaceability system as the basis for negotiations in the Interim Assessment Process. It was also necessary to take the Commonwealth criteria of July, 1995 and refine them further so that conservation goals could be framed precisely in terms of numbers of hectares or occurrences of forest types and species.

The process of policy development that had begun with broad principles in 1992 culminated, in April and May, 1996, in four weeks of intensive negotiations over new conservation areas in production forests. The negotiators were representatives of key agencies and lobby groups with interests in forest management in eastern New South Wales. Once the decision had been made to approach the problem of forestry reform through an explicit negotiation process driven by quantitative conservation goals, a great deal of preparation was necessary if the negotiations were to succeed. Some of these preparations were technical, but a crucial factor in the success of the exercise was the handling of the views and concerns of the parties that would take part in the exercise.

THE LEAD-UP TO THE NEGOTIATIONS: RULES OF ENGAGEMENT, DATASETS, CONSERVATION TARGETS AND FURTHER SOFTWARE DEVELOPMENT

The contentious nature of the forestry debate required not only the involvement of stakeholders in negotiations over the location and extent of new conservation areas, but their involvement in decisions on the goals, methods and results of the Interim Assessment Process. This part of the preparation was coordinated by the Resource and Conservation Assessment Council. It included discussions over how the process would be approached and managed, technical briefings, demonstrations of the interactive system, the establishment of conservation and socio-economic working groups, and development of guidelines for identifying features to be targeted for conservation and the specific targets given to each feature. It involved major background studies on the social and economic implications of deferring areas of production forest from logging and required consultation and information sessions with local communities in the regions affected. Details on these and other aspects of the preparation for the Interim Assessment Process are provided in a report (RACAC 1996).

Before the negotiations, it was agreed that the planning units for the process would be logging compartments averaging about 250 ha in size. It was also agreed that the 2.4 million ha of forests covered by the Interim Assessment Process would be subdivided into eleven regions. The regions were amalgamations of forestry management areas and were chosen to produce manageable numbers of compartments for computer analysis and interpretation by negotiators. Irreplaceability analysis would allow each of the compartments to be assessed in a regional context.

There would be seven stakeholders participating in the negotiations: the Resource and Conservation Assessment Council (also acting as referee in the case of disputes), the forestry industry (the Forest Products Association), the forestry workers union (the Construction, Forestry, Mining and Energy Union), State Forests of New South Wales, the National Parks and Wildlife Service, the conservation movement, and the Commonwealth. The process would continue if any party walked out of the negotiations. A geographic information system was made available to stakeholders and their specialists and advisors before the negotiations so that they could scrutinise the data produced by other parties and complete their own background analyses. It was also accepted that the process would be based on data that were inevitably less than perfect in terms of coverage and accuracy, but all parties acknowledged that the information was the best possible in the time available. The result of these preparations as well as agreement on scenarios for each region and consultation on conservation targets, below, was a commitment by all the stakeholders to see the process through.

It was recognized early in the planning phase that agreement by all parties on a single scenario (a balance of conservation and logging areas) for any region would be very difficult, if not impossible. It was therefore decided that up to four scenarios would be developed by the negotiators for each region. One was the conservation scenario, which was expected to achieve all or most conservation targets. The other scenarios were to make available about 30%, 50% and 70% of the 1995 quota sawlog volumes. The purpose of the negotiations was therefore to decide, for each scenario and for each region, on the configuration of compartments that were to be deferred from or still available for logging.

An important aspect of preparation was to decide on which features would be targeted for conservation, how data on the distribution of those features would be derived, and how much area or how many localities of each feature were to be covered by conservation areas. This led to the development of datasets on forest types in each negotiation region, including estimates of their pre-European extent, new mapping of old growth and rainforest, identification of wilderness, and additional surveys of flora and fauna. Conservation targets for each forest type were based initially on the Commonwealth benchmark of 15% of its pre-European extent (if targets are framed in terms of remaining forest areas, the percentage conserved can be increased by reducing the remaining extent of forest). This benchmark figure was adjusted according to natural rarity, past depletion and vulnerability to clearing to give each forest type a specific target area. Another system was used to define targets for the old growth stages of forest types. Expert panels were established to identify priority species of plants and animals and to decide on species-specific conservation targets. Each panel comprised a representative of the National Parks and Wildlife Service, one from State Forests, and two independents. Following all this work, information had to be compiled on the occurrence of each forest type, including its old growth phase, and each targeted species in roughly 14 000 forestry compartments over the eleven regions. This data base became the raw material for the negotiations.

Once a decision had been made to base the negotiations for the Interim Assessment Process on irreplaceability analysis, the prototype software was taken apart and rebuilt over a period of about nine months. During this time, the developing system was demonstrated to several meetings of stakeholders and was shaped partly by their comments and criticisms. Most of the work went into developing information functions that could tell users about progress towards conservation targets for any feature, the characteristics of any particular compartment, and which compartments had certain combinations of characteristics that would make them suitable or unsuitable for conservation (see Table 1 for a selective list of functions). These functions were designed to make the process as transparent as possible, to allow negotiators to try out alternative goals and ways of achieving them, and to show progress towards conservation targets in a variety of ways. A new statistical approach to estimating irreplaceability was also developed. At the time negotiations began, the conservation software (apart from the geographic information system and data base management system) comprised about 25 000 lines of code, of which only about 300 dealt with the estimation of irreplaceability. The remainder were involved in calculating and displaying information to help the negotiations. About ten person-weeks were spent before the negotiations checking every calculation, testing the display and search functions, and reprogramming to eliminate errors. This was to ensure that all the negotiators could rely on the results and that the process could not be undermined by questions about the validity of the software.

Another line of software development was the wood resources study by State Forests. This was designed to show graphically the timber values of compartments and to identify, from any starting point, the harvesting schedule that would allow a sustained yield of timber over a

Table 1. Examples of information functions for users of the interactive system.

Menu function	Purpose
File | Open Selections	Opens a file that stores the information from an earlier stage of negotiations; this allows all aspects of that earlier stage to be reconstructed, including the screen map and progress towards conservation targets.
File | Edit Stage Memo	Enters a text description of the current stage of negotiations which can be read when deciding which file to open.
File | Autosave	Automatically saves every decision as it is made so that a system crash does not lose information.
Show | Contribution	Graphs the contribution of all areas (e.g., logging compartments) currently deferred from logging (or reserved) in terms of their importance for meeting conservation targets.
Show | Features to Target	Tabulates all the forest types and species in the region showing what percentages of their target areas/numbers have been achieved; also allows areas (e.g., logging compartments) that contain below-target features to be tabulated and mapped.
Show | Selection Log	Shows a text file in which the reason for every decision is logged, together with the area (e.g., logging compartment) for which each decision was made, and the time and date of the decision; search and scanning functions are also available to find particular areas or decisions in large selection logs.
Show | Resource	Shows what percentage of the original timber resource has been deferred from logging (or reserved) or is still available for logging at any stage of the negotiations.
Search | Select As	Allows the user to identify the characteristics of areas (e.g., logging compartments) that should be deferred from logging (or reserved).
Search | Deselect From	Allows the user to identify the characteristics of areas (e.g., logging compartments) that should be taken out of deferral (or intended reservation) and made available for logging.
Search | Map	Allows the user to identify the characteristics of areas (e.g., logging compartments) that should be mapped on the screen for assessment of number, location and configuration.
Report | Irreplaceability	Summarizes the irreplaceability and current status (available for logging, deferred from logging, reserved etc.) of all the areas in the region at any stage of negotiation; exports this information in a format that can be loaded into a spreadsheet package.
Report | Targets	Lists every feature in the region, showing target areas/numbers and what percentage of the target has been achieved at any stage of the negotiation; exports this information in a format that can be loaded into a spreadsheet package.
Options | Restrict Targets	Allows user to choose a subset of all the features in the region (down to individual features if necessary) so that only this subset determines the pattern of irreplaceability.

nominated time frame. This system also featured in the negotiations by providing information on the sustained yield possible under various scenarios for each of the regions.

THE NEGOTIATION PROCESS AND ITS OUTCOMES

The preparation co-ordinated by the Resource and Conservation Assessment Council meant that the negotiations began with many of the contentious issues already resolved. The agencies and lobby groups to be represented in the negotiations were familiar with the aims, methods and rules of the exercise. All the parties had watched the software develop at successive demonstrations, some had helped to shape its development, and all understood the main concepts involved.

The first day of negotiations was April 23, 1996. The venue was the head office of the National Parks and Wildlife Service which filled up with negotiators, their support teams, computer hardware, maps and reports. There were two parallel negotiations in two separate rooms over the following four weeks. One series dealt with the six northern regions, the other with the central and southern regions. Each negotiation room had a principal negotiator from each of the seven stakeholder groups or agencies. Each principal negotiator had one or two assistants or advisors. Also in the room were one or two officers from State Forests to run the wood resources model when this was needed. A key person was the driver of the interactive system who called up information on request, explained functions and calculations, and made suggestions about the best way to analyse data or display the results. Figure 4 is a photograph, obviously taken for publicity purposes, of the principal negotiators for the northern regions grouped around a computer monitor, looking attentively at their supporting hard copy, and smiling benignly. The reality was generally not much different (Fig. 5) — a large table with negotiators spread around four or five monitors so that they could see any information displayed and comment accordingly. For most of the time, the process was as relaxed and cordial as these photographs suggest.

Fig. 4. A publicity shot of the principal negotiators for the northern regions in their negotiation room (photograph by Michael Anderson and courtesy of the Resource and Conservation Assessment Council).

Each of the stakeholders also had a support team with its own work room. The support teams helped to prepare statements on the preferred conservation or industry outcomes for each region, compiled data on individual compartments that was not already in the computerized dataset, and reviewed reports on negotiation sessions. The reports were produced from the conservation software in the form of spreadsheets showing the current status of each compartment in the region (available for logging, reserved, deferred) and how much of the target for each feature had been achieved.

The negotiation process went for four weeks, virtually non-stop. Many of the negotiation sessions extended well into the night, support

Fig. 5. Two views of the actual negotiation process for the northern regions (photographs by Michael Anderson and courtesy of the Resource and Conservation Assessment Council).

teams analysed outcomes from one day and prepared positions for the next, people worked to construct the datasets to be used in forthcoming regions, datasets had to be loaded into the system and checked, preliminary conservation outcomes had to be tested, and reports had to be produced and distributed.

The negotiations for each region began with a position statement from each of the stakeholders, much of which was specific to the region concerned. The first day on each region was usually spent developing the conservation scenario, with industry representatives actively involved in discussions and decisions. In all cases, the conservation scenario failed to meet targets for some features. Representatives of the conservation movement and the National Parks and Wildlife Service recognized that meeting all targets would have had such an impact on the timber industry that the ideal conservation outcome would be politically unacceptable. Achieving all targets would doubtless have led to later pragmatic trimming of deferred areas, probably by people other than the negotiators, which was likely to produce a much less desirable plan than an initially restrained achievement of selected targets. The other three scenarios were generally achieved by "winding back" the conservation scenario in a way that made progressively more timber available, with conservation representatives actively participating.

The number of compartments in regions varied from several hundred to about 3 000. For most regions, the development of the conservation scenario relied initially on automatic search functions to identify compartments with combinations of characteristics that made them valuable for achieving targets. Characteristics used in the searches included the overall irreplaceability of the compartment, its percentage cover of old growth, percentage cover of rainforest, percentage cover by forest types which had still not reached their conservation targets, and the number of species and forest types for which the compartment was highly irreplaceable. Typically, these automated choices were followed by finer scale adjustments to achieve design goals, meet key targets that were still not achieved, and make timber available for the industry where the remaining available volumes were judged to be unacceptably small. The "windback" process took a similar approach of automatic reversal of deferral based on overall timber volumes or volumes of particular species important to industry followed by fine-tuning. The results of each scenario, including the extent to which conservation targets were met for each are summarized by RACAC (1996).

The software performed well, although there were some tense moments on the first day when negotiations for the Eden management area were delayed while a problem with the geographic information system was tracked down and fixed. The negotiations then resumed for about two hours and several people commented that the patterns of irreplaceability seemed to make sense. Then someone questioned the accuracy of the information on a couple of compartments — the species recorded there should not have been present. More problems with the data emerged and the negotiations had to be suspended for the day. Late that night, the problem was tracked down. The system was performing as expected, but there had been a mistake with the preparation of the dataset by people who had been working nights and weekends for some time. The result was that the first day's negotiation for Eden had used random data. The functions for importing data into the system have since been adjusted to make sure that this cannot happen again. The conservation system crashed several times during the four weeks of negotiation, although there were failsafes to prevent loss of information and decisions (Table 1) and the problem was usually fixed within about 30 minutes. The response to a crash was usually relaxed and sometimes jovial, the main issue being the time at which the process would resume.

One of the most important outcomes of the negotiation process, apparent within the first week, was the constructive and mostly relaxed discussion between parties normally at loggerheads over the future of the State's forests. There were frequent out-of-session discussions between former enemies and the goodwill between the participants was especially evident at a party at the end of the four grueling weeks.

AFTER THE NEGOTIATIONS — POLITICAL LOBBYING AND INTERIM DECISIONS

Despite the goodwill, the post-negotiation stage of the Interim Assessment Process, which laid the ground for the final political decisions on the use of large areas of production forest, saw a return to public positions that were often more polarized than those evident during the negotiations. The scenarios for each region were placed on public display in the form of maps, explanatory text and statistics on conservation targets and available timber volumes. Public briefings were also held. Submissions on the outcomes were invited and about 20 000 were received and analysed. These varied from brief letters to major detailed reviews of data and decisions. Comments from the industry side of the debate included criticisms of the Commonwealth criteria, data collection and statistical modelling, calls for minimal reduction in quota volumes, and concerns that the socio-economic impacts of the process were poorly understood. Comments from the conservation side included recommendations for extensive areas of deferred forest to be added to the conservation scenarios with immediate dedication of major new national parks and wilderness.

The analysis of the submissions was presented to the Forestry Subcommittee of the Cabinet Standing Committee on Rural and Natural Resources and there was much independent lobbying by agencies and interest groups, including some not directly involved in the negotiations. The Subcommittee intended to present Cabinet with a single option for each region derived from those negotiated. The development of this option occurred at a high bureaucratic and political level and generally centered on the conservation scenario for each region with changes being made to guarantee short-term availability of timber without unduly compromising conservation values. Industry restructuring packages were also developed as part of the recommended option for each region. Out of this less-than-transparent political process came nine new national parks and nature reserves, about 816 000 ha of deferred forest, extensive new wilderness areas, and agreements on the supply of hardwood for five years.

FUTURE DIRECTIONS

The Interim Assessment Process was the first exercise in conservation planning in New South Wales, and in a wider context, to be based on explicit goals and strategic decisions across regions. It was also the first exercise in the State to openly involve a range of agencies and interest groups in a negotiation process, one that moreover proved itself to be workable. It was actively and publicly supported at senior political levels and has set an important precedent for New South Wales as well as providing a model for conservation planning elsewhere.

The process was focused on the eastern forests, the most charismatic and politically sensitive environment in New South Wales. This was to the detriment of other parts of the State that are less charismatic but arguably more urgently in need of conservation action. The wheatbelt and western slopes are good examples (Benson 1991; Sivertsen 1994). Nonetheless, the Interim Assessment Process has benefited these regions indirectly, first by greatly strengthening the case for explicit regional assessments rather than piecemeal decisions on land use, and second, by providing the technological and political methods for such assessments. Two regional planning exercises in western New South Wales covering the Riverina and Cobar Peneplain biogeographic regions (see Thackway and Cresswell 1995 for boundaries) are now using the new software developed for the Interim Assessment.

Agencies and lobby groups are currently involved in preparations for the Comprehensive Regional Assessments which will extend over the next two years or so. These assessments will be more difficult and controversial for two main reasons. First, they will be much more extensive geographically, dealing with private and leasehold lands and covering forests throughout the State. Second, the implications will not be interim, but long term. The level of participation of interest and community groups will be expanded by establishing regional advisory committees and the number of stakeholders will increase, notably to include Aboriginal and mining interests.

Solutions to problems arising in the Interim Assessment Process will be developed for the next assessments. Once the negotiators in the Interim Assessment became used to the conservation software, they began to identify its limitations, mostly in relation to processing speed and presentation of information. The system is currently being refined further for both speed and flexibility in the way it provides information to users. The wood resources modelling by State Forests is also being refined. Work is underway to prepare new datasets which will contain more survey sites, be more rigorously validated, and include more information on predicted distributions and habitat requirements of species. Most importantly, work is underway to streamline the integration of conservation and socio-economic values for two main parts of the Comprehensive Regional Assessments — the negotiation phase during which options for land use will be developed and the phase in which the options are weighed for final political decisions on the outcomes.

A better integration process for the Comprehensive Regional Assessments could improve the process of forest planning in several important ways. A group is presently being formed to work through the integration issues and the options for solving the problems that became evident in the Interim Assessment or that are now emerging for the comprehensive process. Four of the areas that need improvement are listed below:

1. It will be important to have broad agreement on the relative priority for conservation

management of forest types and species in the likely event that the conservation targets for some will have to be compromised. In the Interim Assessment, practical constraints on available areas for conservation, even for the conservation scenarios, forced decisions during negotiations on which features should be brought up to target and which should not. Without any broad discussion or agreed rationale, these decisions were inevitably *ad hoc*.

2. There is much scope for streamlining the communication between the conservation and timber resource software and making conflicts and trade-offs more obvious during the negotiations.

3. Measures need to be developed to summarize the conservation and socio-economic outcomes of each of the scenarios produced during the negotiations. Conservation scenarios from the negotiations for the Interim Assessment were fine-tuned to produce recommendations to Cabinet, and the rationale for the subsequent adjustments and for the locations of the nine new reserves was not altogether obvious. Improved techniques for comparing scenarios will involve broadly agreed criteria for measuring progress in the development of systems of protected areas. These will help to ensure that the final political and bureaucratic decisions are accountable and explicit.

4. An explicit, quantitative approach is needed for measuring conservation priority, especially in relation to the relative timing of conservation efforts. Progress towards such an approach is underway (Pressey 1997). This new measure will be particularly important for the forthcoming assessments which will include private and leasehold lands. It will focus attention on real regional priorities for conservation action in contrast to the traditional concentration on the public lands easiest to reserve but least in need of reservation (Pressey 1995).

ACKNOWLEDGEMENTS

The development of the conservation software used in the Interim Assessment Process relied almost totally on work by Matthew Watts. Simon Ferrier also made important conceptual and technical contributions. The maps for this chapter were prepared by Giselle Whish, Matthew Watts and Katrina McKay. Constructive comments on the draft, particularly on the later sections, came from Leanne Wallace, Jo White and Tim Shepherd.

REFERENCES

Anon., 1992. National Forest Policy Statement: a New Focus for Australia's Forests. Australian Government Publishing Service, Canberra.

Anon., 1995. National Forest Conservation Reserves: Commonwealth Proposed Criteria. Commonwealth of Australia, Canberra.

Benson, J., 1991. The effect of 200 years of European settlement on the vegetation and flora of New South Wales. *Cunninghamia* **2:** 343–70.

Bonyhady, T., 1993. Places Worth Keeping: Conservationists, Politics and Law. Allen and Unwin, Sydney.

Dargavel, J., 1995. Fashioning Australia's Forests. Oxford University Press, Melbourne.

Franklin, J. F., 1993. Preserving biodiversity: species, ecosystems, or landscapes? *Ecol. Appl.* **3:** 202–05.

Grumbine, R. E., 1994. What is ecosystem management? *Cons. Biol.* **8:** 27–38.

Lunney, D. and Leary, T., 1988. The impact on native mammals of land-use changes and exotic species in the Bega district, New South Wales, since settlement. *Aust. J. Ecol.* **13:** 67–92.

Lunney, D. and Moon, C., 1988. An ecological view of the history of logging and fire in Mumbulla State Forest on the south coast of New South Wales. Pp. 23–61 *in* Australia's Ever Changing Forests ed by K. J. Frawley and N. M. Semple. Australian Defence Force Academy Department of Geography and Oceanography Special Publication Number 1, Canberra.

Miller, K. R., 1996. Balancing the Scales: Guidelines for Increasing Biodiversity's Chances through Bioregional Management. World Resources Institute, Washington DC.

Pressey, R. L., 1994. *Ad hoc* reservations: forward or backward steps in developing representative reserve systems? *Cons. Biol.* **8:** 662–68.

Pressey, R. L., 1995. Conservation reserves in New South Wales: crown jewels or leftovers? *Search* **26:** 47–51.

Pressey, R. L., 1997. Priority conservation areas: towards an operational definition for regional assessments. Pp. 337–57 *in* Selection and Delimitation of Parks and Protected Areas ed by J. Pigram and R. Sundell. Centre for Water Policy Research, University of New England Press, Armidale.

Pressey, R. L., Ferrier, S., Hutchinson, C. D., Sivertsen, D. P. and Manion, G., 1995. Planning for negotiation: using an interactive geographic information system to explore alternative protected area networks. Pp. 23–33 *in* Nature Conservation 4 — the Role of Networks ed by D. A. Saunders, J. L. Craig and E. M. Mattiske. Surrey Beatty & Sons, Chipping Norton.

Pressey, R. L., Humphries, C. J., Margules, C. R., Vane-Wright, R. I. and Williams, P. H., 1993. Beyond opportunism: key principles for systematic reserve selection. *Trends Ecol. Evol.* **8:** 124–28.

Pressey, R. L., Johnson, I. R. and Wilson, P. D., 1994. Shades of irreplaceability: towards a measure of the contribution of sites to a reservation goal. *Biodiv. Cons.* **3:** 242–62.

Pressey, R. L. and Tully, S. L., 1994. The cost of *ad hoc* reservation: a case study in western New South Wales. *Aust. J. Ecol.* **19:** 375–84.

RACAC (Resource and Conservation Assessment Council), 1996. Draft Interim Forestry Assessment Report. RACAC, Sydney.

Richards, B. N., Bridges, R. G., Curtin, R. A., Nix, H. A., Shepherd, K. R. and Turner, J., 1990. Biological Conservation of the South-East Forests — Report of the Joint Scientific Committee. Australian Government Publishing Service, Canberra.

Sivertsen, D., 1994. The native vegetation crisis in the wheat belt of NSW. *Search* **25:** 5–8.

Taplin, R., 1992. Adversary procedures and expertise: the Terania Creek Inquiry. Pp. 156–82 *in* Australian Environmental Policy: Ten Case Studies ed by K. J. Walker. New South Wales University Press, Sydney.

Thackway, R. and Cresswell, I. D., 1995. An Interim Biogeographic Regionalisation for Australia. Australian Nature Conservation Agency, Canberra.

Watson, I., 1990. Fighting over the Forests. Allen and Unwin, Sydney.

World Resources Institute, 1992. Global Biodiversity Strategy. WRI, Washington DC.

CHAPTER SEVENTEEN

Burning issues: communicating fire research in northern Australia

ALAN N. ANDERSEN[1] and BARBARA J. McKAIGE[1]

Fire is certainly a "burning" issue in the savannahs of northern Australia. In many regions, more than half of the landscape is burnt annually, and there is widespread public confusion as to why this is so. This is compounded by a poor scientific understanding of the ecological effects of fire. In 1988, CSIRO Wildlife and Ecology established a landscape-scale fire experiment at Kapalga Research Station in Kakadu National Park, Northern Territory. The major communication challenges have been: (1) Overcoming the general public's perception of fire as a catastrophic event, rather than as a useful tool for land management; (2) Achieving ownership of the research findings by the project's major stakeholders — north Australian fire managers; and (3) Convincing Aboriginal traditional land owners of the value of our scientific research. Communication activities directed at the public have been generally successful, but there have been substantial difficulties in developing ownership of the research by Park managers and traditional Aboriginal land owners. Factors contributing to these difficulties include inadequate consultation during research planning, and the different institutional and social cultures of stakeholder groups. Important lessons we have learned for effective communication are: (1) Effective communication is built on effective stakeholder consultation at all stages of research planning, the conduct of the research itself, and the extension of its findings; (2) Different stakeholders have different needs, and information should be appropriately targeted; and (3) Communication activities require ongoing evaluation of their effectiveness.

INTRODUCTION

FIRE is used in the management of fire-prone ecosystems throughout the world (Whelan 1995; Pyne *et al.* 1996), but is an especially prominent tool for land management in Australia. The traditional use of fire by Aborigines over the past 40 000 (or more) years has helped shape the contemporary Australian landscape (Horton 1982; Pyne 1991), and, since European settlement, anthropogenic fire has continued to play an important role in land management (Gill 1977). The Australian population is heavily concentrated in the southeastern coastal fringe, where fuel-reduction burns aimed at preventing catastrophic forest fires are a familiar, and highly controversial, practice (Gallus 1994; Williams and Gill 1995; Calver *et al.* 1996). Fire behaviour is markedly different in northern Australia, but its management is an equally contentious issue (Morton and Andrew 1987).

The dominant vegetation type of northern Australia is tropical savannah, featuring a dense grass layer occurring in association with sparse trees (Gillison 1983; Mott *et al.* 1985), as is the case throughout the tropical world wherever rainfall is highly seasonal (Boulière 1983). The profuse grass production of the wet season becomes tinder dry over the ensuing dry season, making fire an annual occurrence in savannah biomes (Gillon 1983). Most savannash elsewhere in the world are highly modified by people, and fires are lit primarily in the context of land clearing and livestock production (Gillon 1983). Australian savannahs, on the other hand, are largely uncleared, and fires are lit primarily in a conservation or Aboriginal land management context (Russell-Smith 1995; Andersen 1996).

Fire behaviour varies markedly throughout the northern Australian dry season. Fires lit early (up to June) in the season, when the vegetation is still moist, tend to be very low in intensity, patchy, and limited in extent. Late-season fires, on the other hand, often completely incinerate ground-layer vegetation, cause substantial leaf scorch in the canopy, and cover large areas (Braithwaite and Estbergs 1985; Russell-Smith *et al.* 1997; Williams *et al.*, in press). The dominant management paradigm is one of extensive early season burning, aiming "to create a mosaic of burnt and unburnt country, thereby maintaining habitat diversity and reducing the risk of widespread, destructive late dry season fires" (Salmon 1992). It needs to be noted, however, that even the most severe savannah fires are only of moderate intensity by southern Australian standards, with the canopy conflagrations of southern forests never occurring, even after the long-term absence of fire (Williams *et al.*, in press).

[1]Wildlife and Ecology, CSIRO Tropical Ecosystems Research Centre, PMB 44 Winnellie, Northern Territory, Australia 0822.

Pp. 88–96 *in* ECOLOGY FOR EVERYONE: COMMUNICATING ECOLOGY TO SCIENTISTS, THE PUBLIC AND THE POLITICIANS ed by R. T. Wills and R. J. Hobbs. Surrey Beatty & Sons, 1998.

Kakadu National Park, in the Top End of the Northern Territory (NT), is the premier nature reserve of Australia's seasonal tropics, and fire plays an integral role in its conservation management (Walker *et al.* 1985; Press 1987; Russell-Smith 1995). Fire, however, is also a major public relations concern in Kakadu, with widespread public discontent occurring over what is perceived as unnecessarily frequent and extensive burning. An important factor contributing to the controversy is inadequate scientific information on the ecological effects of fire, and in this context CSIRO Wildlife and Ecology in Darwin (subsequently referred to as DWE) recently completed a large-scale fire experiment at Kapalga Research Station in Kakadu. This chapter discusses the challenges we have experienced in the communication of the findings from the experiment. We believe the lessons we have learned are broadly relevant to other ecological research programmes directed at land management issues.

FIRE IN KAKADU — THE KAPALGA EXPERIMENT

About half or more of Kakadu is burnt each dry season (Braithwaite and Estbergs 1985; Press 1988; Russell-Smith *et al.* 1997), and this coincides with a heavy concentration of tourist visitation. These visitors are lured to Kakadu by advertisements featuring vast wetlands, teeming wildlife and spectacular waterfalls. The reality of a countryside burning apparently out of control, blackened landscapes and smoky skies can be a major shock! This initial shock is exacerbated by widespread confusion as to why there is so much burning, which is compounded by inadequate information on the ecological effects of fire.

Our work at Kapalga was aimed at redressing the lack of information. We were concerned that much conventional wisdom about fire in the region was based on visual impressions, personal experience and opinion, rather than from rigorous scientific study, and that many fundamental ecological issues were being largely ignored (Andersen and Braithwaite 1992; Andersen 1996). We also wanted to avoid the common problem of small plot size that limits the usefulness of most fire experiments. Such an experiment had already been established at Munmarlary, elsewhere in Kakadu (Hoare *et al.* 1980); although it had yielded valuable information on the responses of particular plant and animal groups, it was inappropriate for studies of larger scale ecological patterns and processes (Andersen and Braithwaite 1992; Andersen 1996).

The Kapalga experiment (Andersen *et al.* 1998) was established in 1988, and used 15–20 km^2 water catchments as experimental units. Four burning regimes were applied, representing a range of fire types occurring in the region: (1) "Early" — fires lit early in the dry season, which is the predominant management regime in the region; (2) "Late" — fires lit late in the dry season, as occurs extensively as unmanaged "wildfires"; (3) "Progressive" — fires lit progressively through the dry season, such that different parts of the landscape are successively burnt as they dry out (we understood this to approximate traditional Aboriginal burning practices; Braithwaite 1995a); and (4) "Unburnt" — no fires lit, and wildfires excluded. After the collection of baseline data, all burning treatments were applied annually from 1990 to 1994. Extensive measurements were taken of fire behaviour, and major studies were conducted on the responses of nutrients and atmospheric chemistry, temporary streams, vegetation, insects and other arthropods, small mammals, and vertebrate predators to the burning treatments (Andersen *et al.* 1998).

MAJOR STAKEHOLDERS

Aside from the scientific community, which we are not considering here, we identified three major stakeholder groups for the Kapalga experiment:

1. *The general public.* The Top End of the NT is extremely sparsely populated, with only about 100 000 people living in the entire region. However, Kakadu receives approximately a quarter of a million visitors each year, and has possibly the highest public profile of any conservation reserve in Australia. For example, the proposed mining of Coronation Hill in what is now Stage III of Kakadu was one of Australia's major conservation controversies of the past decade (Resource Assessment Commission 1991), and ongoing uranium mining in the region (Unger and Milnes 1992) continues to attract intense national attention.

2. *Regional fire managers.* Kakadu is managed by a federal organization, Parks Australia (previously the Australian Nature Conservation Agency), in a joint arrangement with the traditional Aboriginal land owners (see below). Throughout most of the rest of the region, including all other conservation reserves, fire management is the responsibility of the NT Bushfires Council, operating within the NT Government's Parks and Wildlife Commission.

3. *Aboriginal traditional owners.* Aboriginal people are major resident stakeholders in Australia's savannahs (Pearce *et al.* 1996). Approximately 50% of the NT is Aboriginal land under

the *Aboriginal Land Rights Act (1976),* and traditional fire management is an extremely important issue for Aboriginal people (D. Rose 1995). Much of Kakadu is Aboriginal land under the Act, and Aboriginal traditional owners have majority membership (represented by 10 of 14 positions) of the Kakadu Board of Management, Kakadu's executive policy-setting body. Parks Australia has a policy of managing all of Kakadu as Aboriginal land, although the western section (Stage II, which includes Kapalga) is currently not (Kakadu Board of Management 1991). There are no longer living descendants of the original Kunbudj traditional Aboriginal owners of Kapalga (Braithwaite 1995b; P. Wellings, pers. comm.), but Aboriginal custom is that there is succession to the land by adjacent clans in such circumstances.

These stakeholder groups represent divergent cultures, values and institutional structures, and have provided a wide range of communication challenges.

COMMUNICATING WITH THE PUBLIC

Most visitors to Kakadu arrive with the temperate experience of catastrophic fires, and view fire as a fearful and destructive agent. The savannah reality is that frequent fire is a normal part of the annual wet-dry cycle, and it is a question of when, not if, fire will occur. The manipulation of fire timing, intensity and frequency then becomes a potent tool for conservation management at the landscape scale. Communication with the public therefore involves a cultural shift from fire suppression to active use of fire for habitat management. Our experience is that members of the general public have a great thirst for information, and the prime communication challenge with them has been to package such information as effectively as possible.

Our communication activities directed at the general public have included information leaflets, public displays and seminars, numerous appearances in the electronic media, and in-service training of primary and high school teachers (Table 1). We have a Consultative Committee comprised of representatives of local industry, conservation and Aboriginal groups, and this committee plays a strategic role in the facilitation of the programme's communication effort. The committee has provided valuable advice on the sort of information required by the public, and the most effective ways of presenting it. For example, before preparing information leaflets based on our preliminary research findings (Table 1, Fig. 1), we asked committee members to consult their constituents on the most common questions asked about fire ecology and management, which would then form the basis of the information contained in our

Table 1. Fire communication activities by DWE Darwin staff, targeted at the general public.

Communication Tactic	Description Of Activities
Media	>60 radio interviews plus five ABCTV news items, two current affairs stories on the ABC 7.30 Report, and magazine-style stories on Quantum and Beyond 2000 in past 5 years.
Information leaflets	Summary information leaflets, which have been distributed to : • tourists via the Darwin Regional Tourist Association office and Gagudju Crocodile Hotel in Kakadu. • the general public via presentations to community groups, the Darwin Show, public libraries and in response to requests for information. • students via school libraries and inservice teacher training courses (see below).
Publications	An extended booklet has been produced on the Kapalga Fire Experiment, containing a bibliography and detailed information about Kapalga Research Station, projects and facilities. The booklet has been updated annually.
Displays	Kakadu Visitors Centre (1991–1994), the foyer of the Park's premier hotel (1991 onwards), and at the Darwin Show (1996).
Public talks	>25 talks on Top End fire ecology and management to community groups, TAFE students, tour guides, and the general public.
Teacher training	Fire ecology is included in annual courses (running for the past 6 years) on Top End ecology, as part of in-service training for primary and secondary school teachers. The courses include a field component, and are formally evaluated for their communication effectiveness.
Kapalga Open Day	More than 160 people attended an open day at Kapalga in May, 1995, commemorating DWE's 20 year involvement with the site, and featuring the fire experiment. A range of information material was produced, including booklets, scientific posters, and a video demonstrating the experimental burning. Guided tours of the fire experiment were conducted by DWE staff. An evaluation indicated that holding an open day was a highly effective way of communicating fire research.

BURNING ISSUES

FOR CONSERVATION IN THE TOP END

LEAFLET No.1

Fire in the Top End

Fire is certainly a 'burning' issue in the Top End of the Northern Territory! Every Dry Season (May-October) the countryside is ablaze, and people's passions are likewise inflamed. Most fires are deliberately lit, and there is widespread public confusion as to why this is so. The constant palls of smoke are a concern to the health of residents, tourists express dismay at the charred landscapes, and scientists debate the effects of the fires on the Top End environment.

Why are there so many fires in the Top End?

Fires have been burning the bush in the Top End for tens of thousands of years. Each year Wet Season rains produce lush plant growth. At the end of the Wet, the vegetation quickly dries out, providing a large amount of fuel for Dry Season fires. Burning in the Top End is carried out for a variety of reasons :

- management of conservation areas, such as Kakadu National Park.
- protection of property
- improving pasture for cattle production
- land management by Aboriginal traditional owners.
- protection of fire-sensitive vegetation, such as monsoon rainforest.

Even if people didn't light fires, the high incidence of lightning strikes during the build-up to the Wet Season will always ensure that fire is a regular event. **The burning question is *when*, not if, fire will occur.**

When is burning carried out?

Most fires are lit by land managers early during the Dry Season (May-June), when the ground-vegetation is still moist. Fires at this time of the year tend to be very low in intensity, patchy, and limited in extent. Such fires remove fuel from the country that is burnt, and so restrict the spread of higher intensity fires which may occur later in the year. This controlled burning is not done to prevent the Ash Wednesday-type crown fires (where flames carry long distances from tree to tree within the canopy) that occur in southern Australia. Such crown fires never occur in the Top End, even if forests are not burnt for many years.

Have burning patterns changed?

There is considerable debate over the extent to which burning patterns have changed in the Top End since European settlement. Unfortunately, details of the timing and extent of Aboriginal fires have been poorly of documented, so this issue may never be resolved. However, there is a common perception that the extent of high intensity fires late in the Dry Season has increased since the disruption of traditional Aboriginal burning practices, particularly in the sandstone country of western Arnhem Land.

Fire is a natural part of the Top End environment

Visitors to the Top End often bring with them a fear of fire, and see it as a destructive element. However, descriptions of burnt land as 'destroyed' or 'devasted' are emotive, and do not reflect the true role played by fire in the balance of nature. Fire is very much a part of the Top End environment. The challenge is to gain a better understanding of it, and to manage it wisely.

Fire Information

This leaflet is the first in a series on fire in the Top End, based on a large fire experiment CSIRO has completed at Kapalga Research Station in Kakadu National Park. The Kapalga Fire Experiment is outlined in leaflet No.2.

Other leaflets include:
3. Up in smoke! Fire and the Atmosphere
4. Fire and Plant Nutrients
5. Fire and Streams
6. Fire and Plants
7. Fire and Animals

FURTHER INFORMATION:
The Communication Officer
Division of Wildlife & Ecology
CSIRO Tropical Ecosystems Research Centre
PMB 44 Winnellie NT 0821 Australia
ph. (08) 8944 8400
fax. (08) 8944 8444

CSIRO AUSTRALIA

Division of Wildlife & Ecology

The information contained in this CSIRO leaflet is of an advisory nature only. It is provided in good faith and is not claimed to be an exhaustive treatment of the subject.

Fig. 1. Number 1 in our series of information leaflets, aimed at conveying our findings to the general public.

leaflets. The draft text was then reviewed by committee members, who pointed out many unforeseen ambiguities and cases of inappropriate language, which would have promoted antagonism and otherwise reduced the effectiveness of the leaflets.

COMMUNICATING WITH FIRE MANAGERS

Our strategic communication goal was for Parks Australia and the NT Bushfires Council to feel a sense of joint ownership of findings from the Kapalga experiment, and therefore to accept them readily. There has been considerable success with the NT Bushfires Council — a DWE scientist is on the Council, the Council's Executive Officer is a member of our Consultative Committee, the Kapalga experiment has been strongly supported, and findings from it have been enthusiastically received.

We have similarly strong institutional ties with Parks Australia, but, despite having directed many communication activities at Parks Australia staff (Table 2), have not been so successful in having them accept the research and its relevance. This has been due primarily to cultural differences between CSIRO and Parks Australia, and consequent tensions between the two institutions during the planning and conduct of the research. We discuss these issues here in the hope that it will improve understanding of each other's perspectives, and provide lessons for future interactions between research programmes and their land management stakeholders.

1. *Research culture.* CSIRO as an institution has a research culture, where quality scientific research is highly valued even if it has no immediate application. Parks Australia, on the other hand, is an institution with a management culture, where the value of scientific research is directly related to the usefulness of its findings. The majority of Parks Australia staff with on-the-ground responsibility for management have little science background, and few have direct research experience. A common complaint from field staff is that research in Kakadu is driven primarily by external requirements (i.e., the priorities are those of the researchers, and not of Park managers), and that the needs of researchers impose on schedules that are already busy with pressing management issues. The results of research are often a long time coming, and are often presented in forms that are inaccessible to them. Parks Australia has also recognized that it needs to improve its management of research findings (Smyth 1995).

From DEW's perspective, the high national and international status of Kakadu makes it an important site for fundamental research on savannah ecosystems, and such

Table 2. Communication activities targeted at regional fire managers.

Communication Tactic	Description Of Activities
Institutional structures	• DWE representation on the Kakadu Board of Management and the NT Bushfires Council. • Joint DWE and Parks Australia representation on the Kapalga Management Committee. • Assistance by Parks Australia staff in lighting experimental fires. • Parks Australia and NT Bushfires Council representation on DWE Consultative Committee.
Training Courses	Three training courses on Top End ecology and management were held by DWE for Kakadu park rangers.
Newsletters	• DWE contributed to the Kakadu Staff Newsletter for five years (the duration of the fire experiment), with most articles relating to our fire research at Kapalga. • In 1992 DWE Darwin initiated a Newsletter for North Australian Fire Managers.
Field Days	• A field day was held at Kapalga in 1993, as part of a North Australian Fire Managers conference. • Kapalga Open Day (see Table 1).
Direct mail	Research results in the form of briefing notes are routinely mailed to senior Parks Australia and NT Bushfire Council staff.
Formal briefings	Formal briefing sessions are given to fire managers represented on our regional Consultative Committee.
Displays	See Table 1.
Conferences	Kapalga results presented to Bushfire 1993 and 1995 conferences, and 1994 North Australia Fire Managers Workshop. DWE Darwin, in association with the NT Bushfires Council and Parks Australia, is organizing the Bushfire '97 conference in Darwin.

research is required to underpin sound land management. CSIRO is not responsible for short-term research dictated by day-to-day issues. Its scope covers most of northern Australia, and research priorities are set according to national (and international) agendas.

2. *Priority issues.* Whereas CSIROs scientific perspective is that sound ecological research is a priority for effective land management, Parks Australia in reality must manage a diversity of issues. Chief among these are issues relating to Aboriginal rights and land management. Kakadu has broken new ground for conservation management on Aboriginal land (Braithwaite 1992; Braithwaite and Roberts 1995), and its joint management structure has been widely acclaimed. Much of Parks Australia's resources have been directed towards making this management structure work, and to Aboriginal issues in general. Similarly, a considerable portion of Parks Australia's resources is inevitably directed toward the management of visitors, rather than land. In other words, the level of priority given to research by CSIRO is inevitably not shared by Parks Australia.

3. *Adaptive management.* A considerable body of literature has recently developed on "best practice" conservation management (Christensen *et al.* 1996; Mangel *et al.* 1996). Such best practice focuses on the maintenance of ecosystems rather than particular species, and adopts the principles of "adaptive management" (Holling 1978), incorporating clear and unambiguous management goals, careful consideration of the full range of management options, specific plans for monitoring management effectiveness, and clear articulation of the links between science, research and management (Holling 1993; Stanford and Poole 1996). It is our perception that these principles have not been fully adopted in Kakadu, and that this has contributed to the lack of acceptance of research by Park staff.

For example, there appears to be some confusion among Park staff between strategic and operational goals of fire management. The "goal" of lighting fires early in the season to limit late-season wildfires is an operational tactic that is useful only to the extent it serves a strategic goal such as the conservation of biodiversity. If prescriptions are seen as goals rather than tactics, then research becomes largely irrelevant, as it is perfectly clear what is required for successful management. If, on the other hand, prescriptions are seen as tactics in the context of maintaining biodiversity (or other strategic goals), then there is a very clear role for research into the effectiveness of these and other possible tactics. It has been our perception that some Park staff view fire prescriptions as goals rather than tactics, without an appropriate emphasis on monitoring and feedback, and therefore have a poor appreciation of the importance of research to fire management.

4. *Research planning.* CSIRO has often in the past been accused of setting its own research agendas without adequate consultation with its stakeholders. A common paradigm has been for CSIRO to conduct high quality research, then to attempt to convince its stakeholders that its findings are of value to them. This unfortunately was the context in which the Kapalga experiment was established. Parks Australia was frequently "consulted" during planning, but this was more in the form of providing information to Parks Australia's research officers, rather than seeking genuine input from Parks Australia staff. In particular, there was no development of a communication strategy to ensure that the research findings would be adopted by Parks Australia.

5. *Lines of communication.* Throughout the conduct of the experiment we had close dialogue with senior Parks Australia staff, but the outcomes of these discussions were poorly communicated to other Parks Australia staff. Most Park staff were ill-informed about all issues relating to Kapalga, from management arrangements to research findings. Attendances at our training and extension days were often poor, and we were widely regarded as being secretive in our activities at Kapalga.

6. *Management arrangements for Kapalga.* CSIRO has had a research interest in Kapalga since the early 1970s, and began its formal management as a research station in 1976 (Braithwaite 1995b). DWE had therefore had a decade-long association with Kapalga before its incorporation into Kakadu National Park (Stage II) in 1984. We continued to have management responsibility for Kapalga, but effective co-operative arrangements with Parks Australia were not developed until several years later. In the absence of such co-operative arrangements, there was inevitably resentment from some CSIRO staff that we were no longer "masters" of the land, and from some Parks Australia staff that CSIRO was managing a substantial portion of Kakadu for its "exclusive" use. These tensions accentuated the communication problems outlined above.

COMMUNICATING WITH TRADITIONAL ABORIGINAL LAND OWNERS

As with Parks Australia, despite considerable communication effort (Table 3) we have been less than successful with engaging traditional Aboriginal landowners in the Kapalga experiment. Factors contributing to this are as follows:

1. *Access to land.* A protocol for Aboriginal access to Kapalga had been negotiated, but, as with Parks Australia staff, there was a common perception among Kakadu's Aboriginal population that they were unwelcome at Kapalga. Unresolved issues between different neighbouring clans over traditional succession to Kapalga's land also contributed to problems of Aboriginal access. This was a poor basis for effective communication with Aboriginal landowners.
2. *Science culture.* A belief in the scientific method of experimental hypothesis testing is not shared by traditional Aboriginal people, who tend to favour their own "truth" regarding cause and effect in the world, based predominantly on intuitive feelings on what is right (Pearce *et al.* 1996). These feelings are based on their own personal experience in the landscape, and what they have learned as part of Aboriginal tradition. Our experience is that most Aboriginal people are bemused by the systematic pursuit of information that does not have direct management relevance, which is a key ingredient of a science culture. The ancestors of Aboriginal people have been managing fire in their environment for tens of thousands of years, and Kakadu's Aboriginal people have lived all their lives on their land. From an Aboriginal perspective, it is therefore difficult to imagine how western scientists, who do not even live on the land, can tell them anything they do not already know. This lack of trust in science is illustrated by a recent case where Aboriginal traditional land owners refused to give permission for water release from Ranger uranium mine into adjacent Kakadu National Park. Aboriginal people believed that such a release was likely to have serious environmental consequences, despite all independent scientific evidence indicating that this would not be the case.
3. *Traditional knowledge.* Australian science has a poor record of incorporating traditional Aboriginal knowledge in ecological research, and protocols are lacking for engaging Aboriginal land owners in research activity. The recent fauna survey of Uluru National Park is seen as an important milestone in the integration of traditional knowledge and scientific research (Reid *et al.* 1992). Such integration becomes even more complex when research is primarily concerned with ecological processes, where the level of Aboriginal "expertise" is unclear, rather than natural history.
4. *Conservation ethics.* A Central Australian study (B. Rose 1995) has shown that the Western middle-class concept of conservation management is not accepted or even understood by Aboriginal people, and our experience is that this is also the case in Kakadu. Feral animals, for example, are seen by Aboriginal people as having a "rightful" place in the environment, as well as being a valued source of meat. Associated environmental changes are recognized, but are often not perceived as being "unnatural" or considered as representing "degradation". Aboriginal people recognize the rights of feral animals to life, and question the rights of people to deny them this (Deborah Rose,

Table 3. Communication activities targeted at Aboriginal land owners.

Communication Tactics	Description Of Activities
Institutional structures	• DWE membership on the Kakadu Board of Management. • Joint DWE and Aboriginal land owner membership on the Kapalga Management Committee. • Residence of Aboriginal traditional owner at Kapalga.
Training Courses	See Table 2.
Kapalga Open Day	See Table 1.
Direct mail	Information leaflets, briefing notes and research papers about fire ecology and management are routinely mailed to Aboriginal traditional owners.
Displays	Display on fire ecology and management in the foyer of the Park's premier hotel, owned by Aboriginal traditional owners.
Public Seminar	Public talk by Kakadu Park Manager on Aboriginal perspectives on scientific research in Kakadu (May 1994).

pers. comm.). Similarly, Western concepts of fire management for nature conservation are quite different from Aboriginal motives for traditional burning practices.

5. *Priority issues.* When compared with social justice issues relating to land rights, cultural heritage, health, education and employment (Lippmann 1981; Taylor 1993), scientific research is hardly likely to have high priority with Aboriginal people unless it is central to their concerns. In the case of fire management, this is accentuated when research findings have the potential to conflict with traditional practices, and therefore social justice concerns (Braithwaite 1992).

CONCLUSION

The Kapalga fire experiment has a diverse array of stakeholders, representing divergent cultures and value systems. We have put considerable effort into communicating its findings, but, in terms of communication effectiveness, our results have been mixed. There will always be differences in the cultural and value systems of research providers such as CSIRO and their stakeholders, and the challenges these pose for effective communication are formidable. What are the important lessons we have learned for improved communication effectiveness?

1. *Stakeholder consultation.* We consider effective stakeholder consultation to be the key to successful communication. Such consultation is required at all stages of research planning, the conduct of the research itself, and the extension of its findings. We have good systems in place for presenting the research findings (Tables 1–3). However, there is a continuing legacy of inadequate consultation with stakeholders during the planning and conduct of the research. In the Kapalga experiment, we failed to engage important stakeholders in the research, with insufficient attention being paid to ensuring that the research was seen as relevant, was understood and was welcomed. An important lesson in this context is that it is unwise to concentrate communication effort at senior members of stakeholder groups, and expect that the information will then be effectively communicated to other members. "Grass roots" support requires "grass roots" communication.

2. *Targeted communication activities.* We believe it is important to recognize that different stakeholder groups require different research information, presented in different ways. It is essential to identify what each group wants, and how they want it, so that information can be appropriately targeted.

3. *Evaluation of communication effectiveness.* It is easy to assume that our communication activities are working well, simply because they seem right to us. In reality, the issues that are important to us, and the way we like information presented, are often not shared by our stakeholders. Communication activities therefore require ongoing evaluation of their effectiveness.

Stakeholder communication is time-consuming and can be a frustrating business, and like many other aspects of research management can seem a terrible infringement on one's scientific sanctity. However, if ecologists want to make a difference to the way our environment is managed, and to ensure that our work continues to receive public funding, then it must be done and done well. And at the end of the day, if it is any consolation, effective communication with stakeholders is ultimately more rewarding than is writing yet another scientific paper read by an average of 1.2 of our peers!

ACKNOWLEDGEMENTS

We are extremely grateful to Peter Wellings, manager of Kakadu National Park during the Kapalga experiment, for so generously sharing his insights with us — the chapter has benefited tremendously from his input. We also thank Dick Braithwaite, Tracey Churchill, Robert Eager, Tony Hertog, Allen Kearns, Greg Miles, Debbie Rose, Jeremy Russell-Smith, Sarah Ryan, Mark Stafford Smith and Dick Williams for their comments on the manuscript. This is CSIRO Tropical Ecosystem Research Centre publication No. 936.

REFERENCES

Andersen, A. N., 1996. Fire ecology and management. Pp. 179–95 *in* Landscape and Vegetation Ecology of the Kakadu Region, Northern Australia ed by C. M. Finlayson and I. von Oertzen. Klewer Academic Publishers, Holland.

Andersen, A. N. and Braithwaite, R. W., 1992. Burning for conservation of the Top End's savannas. Pp. 117–22 *in* Conservation and Development Issues in Northern Australia ed by I. Moffatt and A. Webb. North Australia Research Unit, Darwin.

Andersen, A. N., Braithwaite, R. W., Cook, G. D., Corbett, L. C., Williams, R. J., Douglas, M., Gill, A. M., Setterfield, S. and Müller, W. J., 1998. Fire research for conservation management in tropical savannas: introducing the Kapalga fire experiment. *Aust. J. Ecol.* **23:** in press.

Bourlière, F. (ed), 1983. Ecosystems of the World 13: Tropical Savannas. Elsevier, Amsterdam.

Braithwaite, R. W., 1992. Black and green. *J. Biogeog.* **19:** 113–16.

Braithwaite, R. W., 1995a. Fire intensity and the maintenance of habitat heterogeneity in a tropical savanna. *CALM Science Suppl.* **4:** 189–96.

Braithwaite, R. W. (ed), 1995b. Kapalga Research Station: Two Decades of Research for the Monsoonal Tropics. CSIRO Division of Wildlife and Ecology, Darwin.

Braithwaite, R. W. and Estbergs, J. A., 1985. Fire pattern and woody vegetation trends in the Alligator Rivers region of northern Australia. Pp. 359–64 *in* Ecology and Management of the Worlds Savannas ed by J. C. Tothill and J. J. Mott. Australian Academy of Science, Canberra.

Braithwaite, R. W. and Roberts, S., 1995. Between bining and balanda: Aboriginal burning and conservation management in Kakadu National Park, tropical Australia. *Wildfire*, September 1995, 29–37.

Calver, M. C., Hobbs, R. J., Horwitz, P. and Main, A. R, 1996. Science, principles and forest management: a response to Abbott and Christiensen. *Aust. For.* **59:** 1–6.

Christensen, N. L. *et al.* (12 other authors), 1996. The report of the Ecological Society of America Committee on the scientific basis for ecosystem management. *Ecol. Appl.* **6:** 665–691.

Gallus, C., 1994. The New South Wales bushfires: developing a national park management strategy. *Search* **25:** 34–6.

Gill, A. M., 1977. Management of fire-prone vegetation for plant species conservation in Australia. *Search* **8:** 20–6.

Gillison, A. N., 1983. Tropical savannas of Australia and the southwest Pacific. Pp. 183–243 in Ecosystems of the World 13: Tropical Savannas ed by F. Bourlière. Elsevier, Amsterdam.

Gillon, C., 1983. The fire problem in tropical savannas. Pp. 617–41 *in* Ecosystems of the World 13: Tropical Savannas ed by F. Bourlière. Elsevier, Amsterdam.

Hoare, J. R. L., Hooper, R. J., Cheney, N. P. and Jacobsen, K. L. S., 1980. A report on the effect of fire in tall open forest and woodland with particular reference to fire management in Kakadu National Park in the Northern Territory. Australian National Parks and Wildlife Service, Canberra.

Holling, C. S. (ed), 1978. Adaptive Environmental Assessment and Management. John Wiley and Sons, New York, USA.

Holling, C. S., 1993. Investing in research for sustainability. *Ecol. Appl.* **3:** 553–5.

Horton, D. R., 1982. The burning question: Aborigines, fire and Australian ecosystems. *Mankind* **13:** 237–51.

Kakadu Board of Management, 1991. Kakadu National Park Plan of Management. Australian National Parks and Wildlife Service, Canberra.

Lippmann, L., 1981. Generations of Resistance: The Aboriginal Struggle for Existence. Longman and Cheshire, Sydney, Australia.

Mangel, M. *et al.* (42 other authors), 1996. Principles for the conservation of wild living resources. *Ecol. Appl.* **6:** 338–62.

Morton, S. R. and Andrew, M. H., 1987. Ecological impact and management of fire in northern Australia. *Search* **18:** 77–82.

Mott, J. J., Williams, J., Andrew, M. H. and Gillison, A. N., 1985. Australian savanna ecosystems. Pp. 56–82 *in* Ecology and Management of the World's Savannas ed by J. C. Tothill and J. J. Mott. Australian Academy of Science, Canberra.

Pearce, P., Jackson, A. and Braithwaite, R., 1996. Aboriginal people of the tropical savannas: resource utilization and conflict resolution. Pp. 88–103 *in* The Future of Tropical Savannas: An Australian Perspective ed by A. Ash.

Press, A. J., 1987. Fire management in Kakadu National Park: the ecological basis for the active use of fire. *Search* **18:** 244–8.

Press, A. J., 1988. Comparisons of the extent of fire in different land management systems in the Top End of the Northern Territory. *Proc. Ecol. Soc. Aust.* **15:** 167–75.

Pyne, S. J., 1991. Burning Bush — A Fire History of Australia. Henry Holt and Co., New York.

Pyne, S. J., Andrews, P. L. and Laven, R. D., 1996. Introduction to Wildland Fire. John Wiley and Sons, New York.

Reid, J., Morton, S. R. and Baker, L., 1992. Traditional Knowledge + Ecological Survey = Better Land Management. *Search* **23:** 249–51.

Resource Assessment Commission, 1991. Kakadu Conservation Zone Inquiry Final Report. Australian Government Publishing Service, Canberra.

Rose, B., 1995. Land management issues: attitudes and perceptions amongst Aboriginal people of central Australia. Central Land Council Cross Cultural Land Management Project.

Rose, D. B. (ed), 1995. Country in Flames. Biodiversity Series, Paper No. 3, DEST Biodiversity Unit and the North Australia Research Unit, Canberra and Darwin.

Russell-Smith, J., 1995. Fire management. Pp. 217–37 *in* Kakadu: Natural and Cultural Heritage and Management ed by T. Press, D. Lea, A. Webb and A. Graham. Australian National University, Canberra.

Russell-Smith, J., Ryan, P. G. and DuRieu, R., 1997. A LANDSATT MSS-derived fire history of Kakadu National Park, monsoonal northern Australia, 1980–1994: seasonal extent, frequency and patchiness. *J. Appl. Ecol.* **34:** 748–66.

Salmon, A. (ed), 1992. Australian National Parks and Wildlife Service Annual Report 1991–92. Australian National Parks and Wildlife Service, Canberra.

Smyth, D., 1995. Coming out bright: an audit of research conducted in Kakadu National Park 1992–1995. Consultancy report to the Australian Nature Conservation Agency, December 1995.

Stanford, J. A. and Poole, G. C., 1996. A protocol for ecosystem management. *Ecol. Appl.* **6:** 741–44.

Taylor, J., 1993. The Relative Status of Indigenous Australians, 1986–1991. ANU Press, Canberra, Australia.

Unger, C. J. and Milnes, A. R., 1992. Rehabilitation at Ranger Uranium Mines. Pp. 221–31 *in* Conservation and Development Issues in Northern Australia ed by I. Moffatt and A. Webb. North Australian Research Unit, Darwin.

Walker, J., Davis, J. R. and Gill, A. M. (eds), 1985. Towards an Expert System for Fire Management at Kakadu National Park. CSIRO Institute of Biological Resources, Division of Water and Land Resources Technical Memorandum 85/2, Canberra. 169pp.

Whelan, R. J., 1995. The Ecology of Fire. Cambridge University Press, Cambridge.

Williams, J. E. and Gill, A. M., 1995. The impact of fire regimes on native forests in eastern New South Wales. Environmental Heritage Monograph Series No. 2. NSW National Parks and Wildlife Service, Hurstville.

Williams, R. J., Gill, A. M. and Moore, P. H. R. (in press). Behaviour of experimental fires in a tropical savanna at Kaplaga, Kakadu National Park, northern Australia. *Int. J. Wildl. Fire.*

CHAPTER EIGHTEEN

Community assistance with koala *Phascolarctos cinereus* sightings from a low density population in the south-west Sydney region.

S. J. WARD[1] and R. L. CLOSE[1]

Detecting koalas *Phascolarctos cinereus* (Goldfuss) in the Campbelltown region, south-west of Sydney, is difficult due to low densities of animals. Sightings by local residents provide valuable information and may lead to the capture of animals for the collection of morphometric data, attachment of ear-tags, and radio-collars. The *Macarthur Advertiser*, a local newspaper, has run a weekly column since October, 1995. This column includes a contact number to report sightings, and provides an avenue for education of the public on koala biology and related ecological issues. Sightings reported by the public have increased more than three-fold since publication of the column, compared to reports collected from 1990 to 1995. Sightings reported prior to publication of the column, observations by University personnel and members of the local National Parks Association tended to be concentrated in the same locations, but sightings reported after publication of the column were widespread. Community sightings have provided valuable information which otherwise would have required intensive search effort. Simultaneously, the programme has provided the community with information about ecological concepts.

INTRODUCTION

RESEARCH on koalas poses logistical problems due to the difficulty of detecting koalas, especially at low densities, and capturing animals in large trees located in rugged terrain. The koala population of south-west Sydney region has an estimated density of only one koala per 10 hectares (Close 1993), and animals often use steep, rocky gorges. This chapter reports the use of community responses to help study the ecology of koalas and overcome the difficulties of studying koalas under such conditions.

BACKGROUND

A number of researchers (Robinson 1978; Reed *et al.* 1990; Smith and Smith 1990; Le Page 1993; Leathley *et al.* 1993; Lunney and Moon 1993; Ridley 1993; Curtin and Lunney 1995; Leathley *et al.* 1995; Moon 1995; O'Brien 1995; Lunney *et al.* 1996; Nattrass and Fiedler 1996; Patterson 1996) have used postal surveys or community reports to obtain information on koala populations (usually distribution data). Smith and Smith (1990) for example, concluded that the koala population in Warringah shire, Sydney, had declined, after reviewing reports of community sightings dating back to the 1940s. We believe, however, that the ongoing publicity for this research, and requests to report koala sightings, are a variation on previously used methods of obtaining information from the community. The approach of using community koala sightings is facilitated by the readily identifiable nature of koalas.

The *Macarthur Advertiser* is a local Fairfax newspaper which is delivered to all homes from Wedderburn to Glenfield (Fig. 1). It has a Circulation Audit Bureau (CAB) distribution of 62 608 (at August 9, 1996). The newspaper is supporting the research through fundraising and a $20,000 donation to provide Steven Ward with a scholarship, and by printing a weekly column written by the two authors. The column includes a contact number to report koala sightings, and provides an avenue for education of the public on koala biology. Not all columns are about koalas; some focus on other species or ecological concepts that are relevant to the south-west Sydney region.

The fundraising is being accomplished through the formation of "Mac's Koala Club" (Fig. 2), as a promotion to raise funds for Steven Ward's scholarship. As part of the promotion a "Mac the Koala" suit has also been created to promote both the Club and the newspaper. This is worn at various community events around the district. This campaign also increases community awareness of the local koala colony.

Prior to 1986 the presence of the koala colony in the area was unknown, except to a few individuals. This knowledge became more

[1]Department of Biological Sciences, University of Western Sydney, Macarthur, P.O. Box 555, Campbelltown, New South Wales, Australia 2560.

Pp. 97–102 *in* ECOLOGY FOR EVERYONE: COMMUNICATING ECOLOGY TO SCIENTISTS, THE PUBLIC AND THE POLITICIANS ed by R. T. Wills and R. J. Hobbs. Surrey Beatty & Sons, 1998.

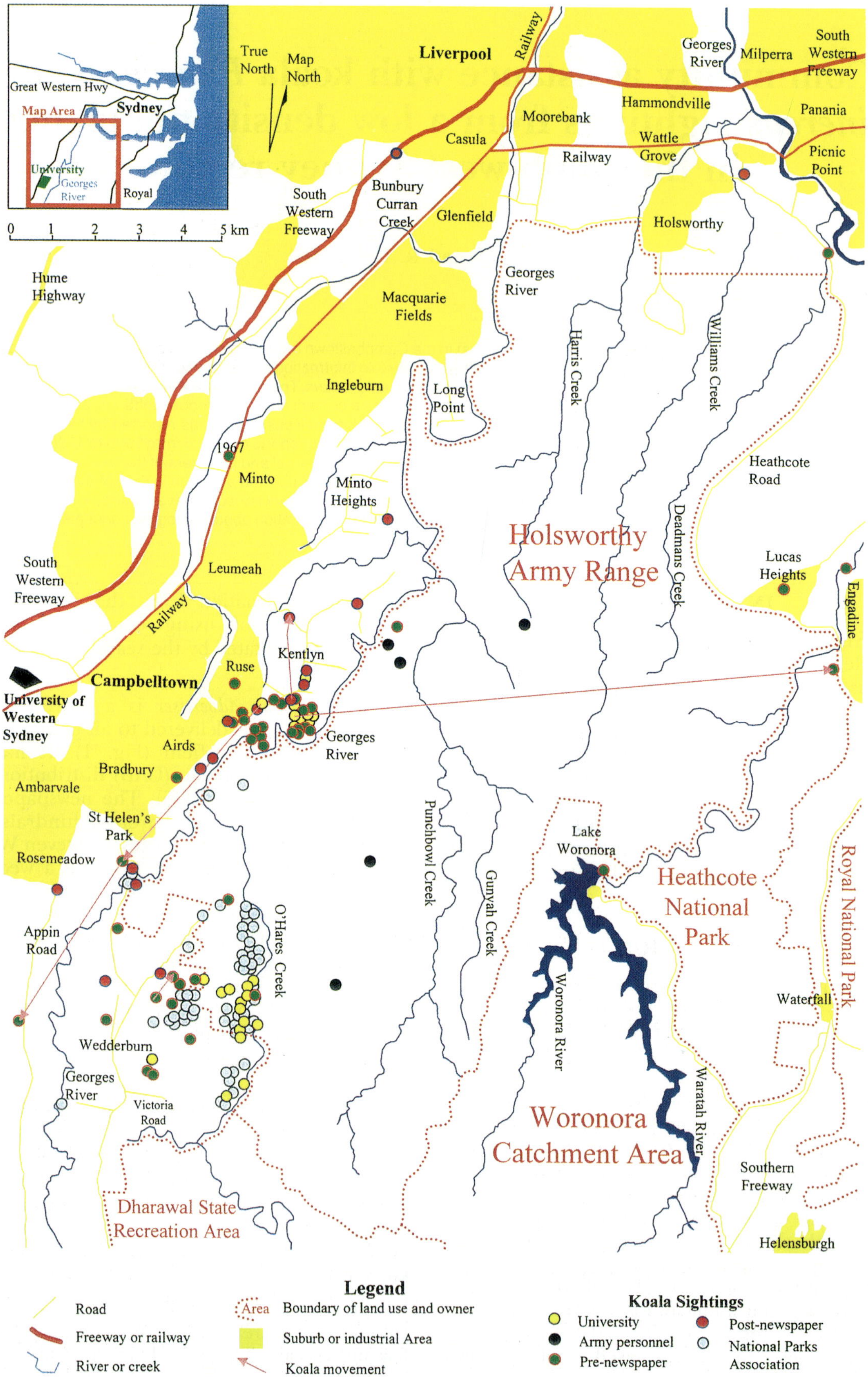

Fig. 1. Regional map (top left) showing area of interest, which is enlarged with koala sightings. Names of some roads, creeks, rivers, and suburbs are shown.

Fig. 2. Form to join Mac's Koala Club, which is published weekly in the *Macarthur Advertiser*. The form printed in the *Macarthur Advertiser* also includes names of members whose birthday occurs during that week.

widely known when a development was approved at the end of Victoria Road in Wedderburn, a suburb to the south of Campbelltown (Fig. 1), and there was considerable publicity about the koala colony at this time. From 1986 to 1990 the members of the Macarthur branch of the National Parks Association noted details of any koalas seen. A CSIRO study of the disputed area (Cork *et al.* 1988) used the sightings available at that time, vegetation data, and observations of pellets or scratches to provide an assessment of the impact of the development on the koala population. As a result of local opposition to the development (Dobson 1990; Sheppard 1990), the construction work was stopped and media attention subsided, although there were intermittent media stories until publication of the weekly column commenced.

Recently the Federal Government has proposed building the second international airport for Sydney in the Holsworthy army range. This area is currently undeveloped, and use of the range has almost exclusively been by army personnel. Shelling of the range, unexploded ordinance, fences, signs along the Georges River, and the rugged nature of the terrain effectively deter most trespassers. An exception is the section of the Holsworthy Army range to the west of O'Hares Creek, which has not been shelled. Access is possible along a fire trail, although human impact so far is minimal.

The south-west Sydney region koala population faces strong pressures from urban expansion and the Holsworthy airport proposal, if it proceeds, is likely to have an exceptionally large impact. Basic ecological data coupled with community understanding of the value of bushland are required if the koala population is to be preserved. This study incorporates both needs.

METHODS

Koalas, when caught, have a coloured ear-tag placed in each ear, and each individual can be identified by its unique colour combination. In all of the *Macarthur Advertiser* columns, readers have been asked to report details of any koala seen and to look carefully for ear-tags; a telephone number is provided in each column.

Sightings for the south-west Sydney region (the area covered in Fig. 1) came from four sources. These sources were: University staff, students and volunteers; Corporal Robert Thompson (Australian Army); National Parks Association records; and public reports made to the University. Sightings reported by the public were classed as being either prior to publication of the newspaper column (pre-newspaper), or after publication (post-newspaper). Multiple sightings reported at the same time were classed as a single report, as were sightings thought to be of the same animal within a week of the first sighting. The time period for the pre-newspaper category was calculated from the first documented report to the day prior to the newspaper column publication, and from the day of publication to the last reported sighting for the post-newspaper category (Table 1). Where possible, sightings have been mapped, but it was not possible to display all sightings in the Wedderburn area (Fig. 1) because of the superposition of symbols.

RESULTS

The average number of sightings per year has increased more than three-fold since the first column was published (Table 1).

The sightings made by the National Parks Association members concentrated strongly in the Wedderburn area. In particular, many occurred near a fire trail extending from Victoria Road, running north near to O'Hares Creek (Fig. 1). The University sightings are also concentrated strongly in Wedderburn, near O'Hares Creek, but with some sightings occurring in Kentlyn (Fig. 1). The sightings reported by Corporal Thompson were few, but widespread within the western section of the Holsworthy Army range (Fig. 1). Pre-newspaper sightings are concentrated in the Ruse, Kentlyn and Wedderburn areas but with reports of sightings at Woronora Dam (to the east of Lake Woronora), the Lucas Heights and Engadine area, the crossing of Heathcote Road over Deadmans creek, and during 1967 in Minto (Fig. 1). Post-newspaper sightings have been widespread, with particularly notable sightings being those at Casula, and to the north-east of Holsworthy (Fig. 1).

Few sightings have been of ear-tagged animals (seven sightings of six koalas to date). Twenty-one koalas have been ear-tagged since 1992, with five of these koalas known to be dead (at August 9). However, three sightings reported prior to the publication of the newspaper column were of two animals which dispersed considerable distances (12.6 and 9.2 km, Fig. 1).

DISCUSSION

It was not possible to establish the veracity of most reported sightings. However, for all investigated sightings the animal was located, or evidence of a koala's presence was found (scratches or faecal pellets).

Following the start of the weekly newspaper column and promotion of Mac's Koala Club the

Table 1. Dates, number of reported koala sightings, and average number of sightings per year for National Parks Association, pre-newspaper (prior to publication of the weekly newspaper column) and post-newspaper (during the period of publication of the newspaper column) categories.

Category	Dates	Number of Reported Sightings	Average Number of Sightings per year
National Parks Association	May 31, 1986–May 4, 1990	66	16. 80
Pre-newspaper	October 31, 1990–October 17, 1995	36	7. 26
Post-newspaper	October 18, 1995–June 24, 1996	16	23. 38

number of koala sightings reported to the University has increased and is probably due to the increased publicity. However, koalas may be coming into locations where people are present (i.e., the suburbs) more frequently, possibly due to development of residential areas, an increase in koala numbers, or a change in their behaviour. Koalas may also be attracted to watered and fertilized trees in suburban gardens, thus resulting in more sightings being reported.

The concentration of University, National Parks Association, and pre-newspaper sightings in the Wedderburn and Kentlyn areas probably reflects the location of most efforts to find koalas. Few sightings have come from the Holsworthy Army range. This probably reflects the small search effort, rather than the true koala distribution. Little work has been conducted to date in the suburbs north or east of the Holsworthy Army range, and the distribution area of the *Macarthur Advertiser* does not extend to these areas (Holsworthy, Moorebank, Wattle Grove).

Twenty-nine per cent of the koalas ear-tagged have been sighted. However, the ear-tags are difficult to see or people may forget to look for them. We hope to obtain more information on koala movements by ear-tagging more koalas, thus increasing the chances of sighting marked animals.

The publication of the column has drawbacks, mainly the time invested in writing. On rare occasions the column has not been printed because of technical difficulties. The Mac's Koala Club membership has steadily increased to over 350, although fundraising has been slow. It is not clear what educational impact the column and Mac's Koala Club have had, but we hope to assess this in the future.

Plans to further enhance the community response include production of a video for distribution to schools in the area, setting up a telephone line which provides a recording of a calling koala male (to allow the public to recognize these calls) and requesting historical sightings. Expansion of Mac's Koala Club to areas with potential koala populations which are served by other Fairfax newspapers is already underway.

This project is not only permitting the study of a species whose low population density would otherwise restrict ecological studies, but should increase the communities' knowledge, and presumably appreciation, of native local flora and fauna. This appreciation is currently being tested by a recent call for submissions concerning the proposed airport in the Holsworthy Army range.

ACKNOWLEDGEMENTS

This research has been supported financially by the *Macarthur Advertiser* and the Australian Koala Foundation. We also thank the *Macarthur Advertiser* for the publicity provided. We thank the Macarthur branch of the National Parks Association, Corporal R. Thompson, and others who reported koala sightings. We thank Wayne Foster and Anthony Scarman for noting or collecting details of koala sightings. The Campbelltown Council and the Australian Army granted access to property.

REFERENCES

Close, R. L., 1993. Campbelltown's koalas: what is their future? *Nat. Parks J.* **37:** 22–25.

Cork, S. J., Margules, C. R. and Braithwaite, L. W., 1988. A survey of koalas and their habitat near Wedderburn NSW, suggestions for management and an assessment of the potential effects of a proposed subdivision of four hectare lots. (A report for Campbelltown City Council). CSIRO Division of Wildlife and Ecology, Canberra.

Curtin, A. and Lunney, D., 1995. A comparison of community-based survey and field-based survey for koalas in a large reserve system on the outskirts of Sydney, New South Wales. Pp. 186–88 *in* Proceedings on a conference on the status of the koala in 1995. Australian Koala Foundation, Brisbane.

Dobson, S., 1990. Can you help us save the koalas at Wedderburn? P. 178 *in* Koala summit. Managing koalas in New South Wales ed by D. Lunney, C. A. Urquhart and P. Reed. NSW National Parks and Wildlife Service, Sydney.

Le Page, C., 1993. Koala preservation society of NSW Inc. report. Pp. 73–74 *in* Proceedings from the conference on the status of the koala in 1993. Australian Koala Foundation, Brisbane.

Leathley, S., O'Connor, S., Lunney, D. and Callaghan, J., 1995. Port Stephens koala management plan. Pp. 120–28 *in* Proceedings on a conference on the status of the koala in 1995. Australian Koala Foundation, Brisbane.

Leathley, S., Lunney, D. and Callaghan, J., 1993. A koala management plan for Port Stephens. Pp. 29–36 in Proceedings from the conference on the status of the koala in 1993. Australian Koala Foundation, Brisbane.

Lunney, D., Moon, C. and Matthews, A., 1996. A 1990 survey of the koala *Phascolarctos cinereus* population at Iluka in northern New South Wales. Pp. 102–22 *in* Koalas — Research for Management. Proceedings of the Brisbane Koala symposium, 22nd–23rd September 1990 ed by G. Gordon. World Koala Research Inc. Brisbane.

Lunney, D. and Moon, C., 1993. Koala surveys. Pp. 41–42 *in* Proceedings from the conference on the status of the koala in 1993. Australian Koala Foundation, Brisbane.

Moon, C., 1995. W.I.R.E.S. koalas tell a story. Pp. 49–54 *in* Proceedings on a conference on the status of the koala in 1995. Australian Koala Foundation, Brisbane.

Nattrass, A. E. O. and Fiedler, K. B., 1996. Koala rescue — the perception and the reality. Pp. 129–36 *in* Koalas — Research for Management. Proceedings of the Brisbane Koala symposium, 22nd–23rd September 1990 ed by G. Gordon. World Koala Research Inc. Brisbane.

O'Brien, D., 1995. Koalas in the Gosford district. Pp. 230–63 *in* Proceedings on a conference on the status of the koala in 1995. Australian Koala Foundation, Brisbane.

Patterson, R., 1996. The distribution of koalas in Queensland — 1986 to 1989. Pp. 75–81 *in* Koalas — Research for Management. Proceedings of the Brisbane Koala symposium, 22nd–23rd September 1990 ed by G. Gordon. World Koala Research Inc. Brisbane.

Reed, P. C., Lunney, D. and Walker, P., 1990. A 1986–87 survey of the koala *Phascolarctos cinereus* (Goldfuss) in New South Wales and an ecological interpretation of its distribution. Pp. 55–74 *in* Biology of the Koala ed by A. K. Lee, K. A. Handasyde and G. D. Sanson. Surrey Beatty & Sons Pty Ltd, Chipping Norton.

Ridley, D., 1993. Koala management in logging areas — an optional managers perspective with reference to Yurammie west forest, Eden district. Pp. 55–73 *in* Proceedings from the conference on the status of the koala in 1993. Australian Koala Foundation, Brisbane.

Robinson, A. C., 1978. The koala in South Australia. Pp. 132–43 *in* The koala ed by T. J. Bergin. John Sands Pty Ltd, Sydney.

Sheppard, J., 1990. The Wedderburn koala colony. Pp. 70–73 *in* Koala summit. Managing koalas in New South Wales ed by D. Lunney, C. A. Urquhart and P. Reed. NSW National Parks and Wildlife Service, Sydney.

Smith, P. and Smith, J., 1990. Decline of the urban koala (*Phascolarctos cinereus*) population in Warringah Shire, Sydney. *Aust. Zool.* **26:** 109–29.

CHAPTER NINETEEN

Ecologists: important links to successful science communication in schools.

J. FISHER[1] and B. CAMPBELL[2]

Schools provide an important focus for the communication of conservation and ecology to the public. City Beach Primary School, its Bush Revegetation Project, Bush Guardians and Conservation and Ecology Programme provides an award winning case study for the involvement of ecologists in school action programmes. This involvement ensures sound ecological principles are used to develop students' appreciation of the natural environment. An ecologist's input enables teachers, who may feel constrained by their lack of knowledge, to be more confident using action science in their teaching. This case study demonstrates how a successful ecological project can provide the impetus for ecology to become integrated into many learning areas.

INTRODUCTION

SCHOOL students' appreciation of the natural environment is enhanced by the incorporation into the school curriculum of action projects, centered around their local environment. Such projects can encourage the development of co-operative behaviour, team work, decision making abilities, problem solving skills and team skills, in both the school and wider community. The case study of City Beach Primary School, to be outlined in this chapter, demonstrates how the knowledge gained by students of their local plants, which in this case are able to tolerate salt, wind, lengthy drought periods and soils of low nutrient, enhances their overall concern for their own environment and their understanding of other environments.

The development, implementation, incorporation and long-term survival of action projects and their supportive programmes within the constraints of school programmes poses many problems. With the school day fully committed to the teaching of the standard curricula, time becomes the major constraint for the inclusion of educational areas which are outside the basic curriculum. A further problem of incorporating action projects into school programmes is that they need and demand expertise not usually available within the school environment. Such projects also require financial assistance not available through school budgetary arrangements. Another difficulty is that the projects must comply with the School Development Plan.

The traditional approach to education using the standard curricula, means a project such as the case study to be outlined here would be an additional educational area over and above those already taught within the school. Additional time would need to be allocated within the teaching week for this programme and additional preparation time would be needed by the staff involved. Under this scenario such projects are only initiated within schools where there is a person with a special interest, willing to devote many extra hours to develop, support and maintain such programmes. In many circumstances these projects are unable to maintain themselves once that person is no longer involved with the school.

Ecology, the relationships between and among organisms and their environment is multidisciplinary, including both descriptive and experimental aspects. Due to its multidisciplinary nature, ecology lends itself admirably to becoming a part of all learning areas within a school programme, incorporating action projects into the existing school programme in many learning areas including English, Maths, Science, Art, and Social Studies. With the provision of expert knowledge by ecologists it becomes possible to organize educational materials across the school and within existing learning areas so that each year level has responsibility for individual aspects of a whole school programme.

Ecologists play an important role in developing students' understandings of their

[1]Department of Botany, The University of Western Australia, Nedlands, Western Australia, Australia 6009.
[2]City Beach Primary School, Marapana Road, City Beach, Western Australia, Australia 6015.

Pp. 103–09 *in* ECOLOGY FOR EVERYONE: COMMUNICATING ECOLOGY TO SCIENTISTS, THE PUBLIC AND THE POLITICIANS ed by R. T. Wills and R. J. Hobbs. Surrey Beatty & Sons, 1998.

local environment by assisting with the establishment and development of projects, implementing research projects, communicating to students and teachers, and curriculum writing in a number of school disciplines. This involvement provides students with scientific investigative and interpretive skills, providing them with an understanding of the processes operating in their environment, and thus the confidence to convey their understandings to the wider community.

The aim of this chapter is to describe City Beach Primary School's successful Conservation and Ecology Programme, which focuses on the natural local environment, has involved ecologists since its inception and has overcome many of the problems inherent in developing school programmes. This has occurred by the integration of action projects, involving the study of plants and animals in their local natural systems, into existing learning areas, enriching the standard formal curriculum with materials the students can easily interpret as part of the real world.

CASE STUDY

City Beach Primary School provides a case study in the involvement of ecologists as links to successful Science communication in schools. The case study to be outlined provides an example of a successful revegetation project and its evolution, which has overcome, or provides strategies to overcome, many of the difficulties inherent in incorporating action projects into school programmes. Since the inception of the Bush Revegetation Project in 1989 the school has benefited from the involvement of ecologists who have interacted with the students and provided expert advice on the many aspects of the project.

DEVELOPMENT PHASE

In 1989 the Parents and Citizens Association resolved to begin a Bush Revegetation Programme. The objective was to revegetate a lawned area with species native to the immediate locality. A draft Management Plan was developed by parent Christina Lake in October 1990, and was circulated to the then Principal, Alan Keane, teaching staff, the Parents and Citizens' Association, local botanist Robert Powell, and Dr Boyd Wykes, with a copy available for parents to view (Lake 1991). Comments received from Robert Powell and Dr Boyd Wykes were incorporated into the Plan. Robert Powell, who became the consultant to the project, provided a list of the local plant species (i.e., those that occur, or used to occur, naturally on the site) to be used to revegetate the site (Lake 1991).

The preface to the Management Plan (Lake 1991) states:

> "It is assumed that staff will incorporate the aims of the project into their teaching whenever possible. The garden will not be large in area, therefore any contribution to the environment made by its establishment will be quite small. It is the students' acquired knowledge of identifying and growing local native plants, and their understanding of the environmental issues concerned, that will make a very large contribution."

These statements have proved to be very accurate when considering the developments which have occurred since the inception of the Revegetation Project.

Scientific advice from experts at this developmental stage ensured that the bush garden would be managed with the aim of achieving four main objectives:

1. It would include as many species as possible from the specified list of plants.
2. It would provide a natural habitat for wildlife.
3. It would resemble the natural bushland in appearance.
4. It would be largely self-sustaining (Lake 1991.)

The provisions and policies of the plan were to operate for four years with a full review to be carried out after this time (Lake 1991).

Initially 108 square metres were cleared in March, 1990, with the first plantings taking place in May, 1990. It was anticipated that the bush garden would be 300 to 400 square metres by 1994. Following the clearing of the site, Dr Kingsley Dixon, Director, Plant Science Division, Kings Park, visited the school and talked to the students about Western Australian plants. He brought a variety of plants to the school, and the students were encouraged to handle them and ask questions.

During the development of the bush garden, coordinating teacher Margaret Park assisted students and staff members. Students were involved in planning and designing the bush garden, digging up plants from a nearby vacant block of land (with permission of the landholder) and planting these and others acquired from APACE WA (Appropriate Technology and Community Environment). Plants obtained from the Local Plants Group excursions were also planted in the bush garden. Permission was obtained for students to collect seeds from Trigg Reserve, a nearby coastal reserve. Seed collection was carried out under the supervision of the consulting botanist.

To conserve the genetic integrity of the bush garden, the origin of plants, seeds and cuttings must be confirmed before being planted. This ensures that they only come from coastal localities within the Perth Metropolitan Region (Lake 1991). Seeds collected by the students were distributed in the bush garden and some were supplied to APACE WA to propagate for planting. To ensure seedlings become as well established as possible before summer, all plantings in the bush garden are carried out in May/June, at the beginning of the wettest part of the year. Some plants needed minimal weekly watering over the first summer, but once plants were established no additional watering was given (Lake 1991).

During this four-year period ecologists from many different organizations visited the school. Some of these included Dr Jonathan Majer (Curtin University), who explained to the students the role played by ants in bushland environments. Some of his students assisted a class to survey and identify the bush garden's ant population. Dr Peter Bindon of the Western Australian Museum explained to students how Aborigines used the plants. The involvement of ecologists has ensured the ecological integrity of the garden and that the school community has been educated with sound ecological principles. Their involvement has provided the school's staff and students with an understanding of interrelationships within their local environment, and a greater appreciation of natural systems.

COMMUNITY INVOLVEMENT

In 1993, at the end of the four-year period of the Management Plan, the project's aims and objectives were assessed. The only aim not achieved was "to provide community awareness of local native vegetation" (Lake 1991).

To fulfill this unrealized aim, and to bring a social dimension to the bush garden, an after-school club was formed in 1994. This was another parent initiative supported by the then Principal, Alan Keane. With the assistance of the Convener, Judy Fisher, and other interested parents, the children's ideas and insights are used to direct the club's activities. The children named their club the "Bush Guardians", and designed a logo which was reproduced on club badges.

During a school Open Day in early 1994, the Bush Guardians were enthusiastic and knowledgeable guides, demonstrating to visitors from the local community the adaptations of local plants to the harsh coastal conditions. Perth has a long dry summer, with rainfall occurring in the winter months. The Bush Guardians' understandings enabled them to explain to visitors the plants' ability to survive without artificial watering, even in periods of severe drought. This highlighted the benefits of local plants to the gardener. Many of the visitors to the bush garden made inquiries about the purchase of these plants, which are not readily available in commercial nurseries. The inquiries gave the students the idea to propagate local plants for the community, with the aim of assisting gardeners to reduce their water consumption. Expert advice in the development of the Nursery and in propagation procedures was gained from APACE WA and technical staff at the Botany Department, University of Western Australia. Funds were needed to establish the Nursery, and the Bush Guardians' idea to collect aluminium cans for this purpose soon saw the development of a community aluminium can collection centre at City Beach Primary School, supported by Westcan. Thirty-five sponsors from the business community supported the development of the Nursery. The Nursery, named the "Plant Gallery" by the students, was established at the end of 1994, with the first seedlings being made available to members of the public in May, 1995, ready for planting prior to the winter rains.

Throughout the development of the revegetation project, the advice of Robert Powell has helped the school to develop a garden mature enough to become a seed orchard for the propagation of local species. The students' knowledge and ecological understandings, gained from their bush garden and interactions with many ecologists, led them to the realization that by propagating the local species for gardeners they were also encouraging them to conserve water.

In the Summer of 1994/1995, strict water restrictions were introduced to Perth and have been in place ever since. Since the introduction of water restrictions a large movement has become established, backed by the Water Corporation, Water and Rivers Commission, Greening Western Australia and Kings Park to educate the public on water conscious gardening. The concept of WaterWise Schools has evolved since this time. The students' ideas and actions, which pre-dated this movement, have received tremendous support from the community, reflected by the large response from members of the public, purchasing local plants for their gardens. Since this time the City Beach community has reduced its water consumption by 14%. Greening Western Australia, the Water Corporation and Alcoa

Australia have supported the initiatives of the students by conducting in-service courses on WaterWise Gardens at the school, during which the Bush Guardians guide the participants around the school's bush garden and Plant Gallery.

Ecologists have played an important role in the Bush Guardians' understandings of their local natural environment, and have enhanced the students learning experiences by providing hands-on interactions, demonstrating the relationships which exist between the local environment and its flora and fauna. Fauna studies have been assisted by Eric McCrum, from Herdsman Lake Wildlife Centre, who has enthralled the students with his demonstrations of the spiders and insects in the school bush garden. John Dell, from the Western Australian Museum, has enhanced studies of the fauna of the nearby Bold Park, a large coastal reserve, by involving the Bush Guardians in catching, identifying and releasing insects. Of great interest to the students has been their involvement with John Dell, in pit trapping for reptiles and his night time spider and possum walks in Bold Park. Shelley James, from the Botany Department, University of Western Australia, has introduced the students to the wonders of insects and plants in the school bush garden via hands on experiences with microscopes.

The Bush Guardians' understandings of their local flora have been extended by studies of plant adaptations to long periods of drought with Judy Fisher, from the Botany Department, University of Western Australia. The students gained further insights into plant adaptations during a walk with Carl Grant, from the Botany Department, University of Western Australia, observing burnt and unburnt bushland following fire in Kings Park. Understandings of the uses of their local plants were enhanced by Noel Morich, (formerly of Kings Park) when he explained the Aboriginal use of plants in the bush garden. Further investigations into plant studies have been observed by the students during a tour of the Botany Department, University of Western Australia, led by Dr Bill Loneragan. To increase their understandings of ecosystem functioning, the Bush Guardians have conducted soil seed bank investigations on the nearby coastal dunes of South City Beach. This is part of preliminary investigations for the development of a Management Plan for a section of the dunes to be managed by the Bush Guardians, City Beach Primary School and the Town of Cambridge, the local council. Shelley James, from the Botany Department, University of Western Australia, has helped the students to develop a plant list of the species present on the dunes.

As all plants grown in the Plant Gallery are local species, it is important that the seed source is local. Vanda Longman, a local botanist, has assisted with seed collection for propagation in the Plant Gallery. Robert Powell has helped the students to propagate native pellitory *(Parietaria debilis)*.

Birds constitute an important component of the local ecosystem and Judy Blyth from the Royal Australasian Ornithologists Union (RAOU) has delighted the students with her bird walks and talks. Dr Boyd Wykes has given the students further understandings of birds in the wild by including them in bird banding sessions in Bold Park.

Another aspect of the students local ecosystem is the wetland. The water quality and mini-beasts of Perry Lakes, a wetland adjacent to Bold Park, have been investigated with the assistance of the Water and Rivers Commission's Ribbons of Blue programme. The Australian Trust for Conservation Volunteers (ATCV) has sponsored a visit to Cockburn Wetlands Centre where ecologists explained the food cycle in a wetland ecosystem.

The development of the Bush Guardians' knowledge and understandings of relationships in the natural world has been enhanced by the involvement of ecologists who have voluntarily contributed their time and expertise. This has provided the students with the confidence to communicate to the wider community. Acknowledgement of the Bush Guardians' communication abilities has been shown by the support of the school, local, scientific and business communities.

SUSTAINABILITY — NATURALLY, FINANCIALLY AND SOCIALLY

The students began recycling aluminium cans to raise funds for the development of a Nursery. With the support of the school community, and the assistance of Westcan, the school became a community collection centre for aluminium cans. The recycling of aluminium cans, the support of the school's 35 sponsors and a minimal charge for the sale of seedlings have financed the school's Nursery, and ensured the school's Conservation and Ecology Programme is now financially self-sustaining. An extension of this has been the collection of aluminium cans, from local beaches, to help fund the school's most recent project, that of working with the Town of Cambridge to restore a sand dune one kilometre from the school. The Plant Gallery is now fully established, comprising a hot house for the propagation of cuttings and two shade houses for propagation by seed. All students in the school have helped to paint a mural which surrounds the Nursery. Artist in residence, Danielle Frichot, assisted with this using the students' illustrations to depict the development of the bush garden. This ongoing community support means all aspects of the school's Conservation and Ecology Programme are now financially secure. Funds from the recycling of aluminium cans, and sale of plants, are used to fund all aspects of the whole school Conservation and Ecology Programme. Just as the school's bush garden is naturally self sustaining, the wider school programme is financially self sustaining and the involvement of the Bush Guardians ensures that all projects are socially sustainable.

CONSERVATION AND ECOLOGY PROGRAMME

The development of the Bush Guardians' club has revitalized interest in the bush garden, and all staff in the school have supported the integration, across many disciplines and throughout the school, of a Conservation and Ecology Programme. As students progress from Years 1 to 7 they have been involved in and are responsible for different aspects of the programme. These aspects have been incorporated into many learning areas, with curriculum materials being written in each learning area to ensure the sustainability of the programme.

Year 1's are propagating the native pellitory plant, the native food plant in southwestern Australia for the Australian Admiral Butterfly. They are also breeding the Admiral Butterfly, *Vanessa itea* (Powell 1993). Using their English skills the students educate other school members about the plant and its interrelationships with the butterfly. Once the plant is mature the students supply them to the school community to plant in their gardens. They propagate sufficient amounts to plant large numbers in the school's bush garden and so attract the larvae of the butterfly to the garden. The propagation of the plant, and the breeding of the butterfly, has been incorporated into the curriculum for English, Mathematics, Science, Art and Studies of Society and the Environment.

To enhance their ecological understandings Year 2 students study mini-beasts and their relationships with the plants in the garden. Ant studies and their relationships with plants, through seed dispersal and harvesting, are the domain of the Year 3 students. Year 4 students are responsible for the monitoring and introduction of fauna into the bush garden. The garden now provides sufficient cover to support small reptiles (Powell 1996). The monitoring is maintained over different seasons, and year by year, to provide an accurate assessment of the changing fauna in the bush garden. The trapping and surveying of the mini beasts, ant communities and reptiles have been incorporated across all learning areas.

Students moving into the senior school, Years 5, 6 and 7, have obtained an appreciation of the importance of interrelationships in the natural world through their studies of the fauna and flora of the school garden. This has occurred with activities which have involved understanding, knowledge and action being incorporated into the traditional learning areas. With this strong background the senior students are in a sound position to tackle their responsibilities. Year 5's Management of the Plant Gallery has been incorporated into all learning areas and provides an introduction to Business Management. The drawing up and implementation of a Management Plan for the restoration of a nearby coastal dune, in association with the Town of Cambridge, is the responsibility of the Year 6 class. All learning areas are being incorporated into this project, including the incorporation of the mapping of the area for weed and native species onto a Geographic Information System. The vital role of communication of all aspects of the School's Conservation and Ecology Programme and Bush Guardian activities is taken on by the Year 7's. They are producers of the school-based radio station, City Beach Radio and are journalists, photographers and editors in the school's quarterly publication "The Bush Press".

All students in the school are involved in the many aspects of the school programmes, and each class has a responsibility towards the overall Conservation and Ecology Programme. By the time students have spent seven years at the school, they will have gained a strong appreciation of ecology and interrelationships in the natural and social world. Parents and ecologists play a vital role, assisting staff members whenever they need support in the classroom or outdoors.

All members of the staff contribute to ensure that these programmes are integrated into the school programme, with one teacher liaising between Bush Guardian and school activities. The Conservation and Ecology Programme is not an additional programme in the school but an integral part of many learning areas. Technology and Enterprise is being incorporated into the Years 5, 6 and 7 responsibilities. Staff at the school are writing curriculum materials to ensure the long term sustainability of the programmes, and to assist other educators who may wish to become involved. The staff have also developed

Curriculum Materials for the WaterWise Programme.

OUTCOMES

The development of the Bush Revegetation Project has provided the school with an educational resource facilitating all learning areas. The bush garden has matured into a bushland area resembling the original vegetation community of the area and is able to support the faunal communities which co-exist with this vegetation. As the garden has matured it has provided the school with a seed orchard and cutting source for the propagation of local species. The involvement of ecologists, providing expert advice throughout the development of the project, and during the many outcomes which have evolved since that time, has ensured that sound ecological principles have been applied in educating students, teachers and community members about their local environment.

The initial outcome from the establishment of the bush garden was the development of the Bush Guardians' Club. This club has played a pivotal role providing students with the means of raising their ecological awareness, and being instigators of award winning environmental projects. The members of this club have educated the community about the value of local species in gardens, both to reduce water consumption and to conserve the nearby bushland of Bold Regional Park. The Bush Guardians have communicated ecological principles to a large audience while acting as Guides on Community Open Days, by guiding visiting teachers and Conservation and Land Management (CALM) employees during a number of inservices held at the school, by being interviewed live for ABC and Commercial Radio, by being the focus of articles in magazines and newspapers, by distributing information via letterbox drops, by developing the Plant Gallery, by creating enthusiasm for projects and programmes throughout the school, and by having fun and being part of an important social group.

The maturity of the garden has been the impetus for the student initiated development of "The Plant Gallery" which has become a community nursery providing local species to members of the public and the Town of Cambridge for revegetation projects. The establishment of this Nursery led to a recycling project, supported by many community members, which has ensured the school's Conservation and Ecology Programme has become self-funding. The establishment of this Nursery has involved a large number of businesses and groups in the sponsorship and development of the Nursery. The bush garden and Nursery have also provided the impetus for close liaison and co-operation with the Town of Cambridge.

Stemming from the establishment of the bush garden, and following the development of the Bush Guardians and Nursery, a further very important development has been the establishment of the school's Conservation and Ecology Programme, which demonstrates a method of integrating a successful ecological project into the constraints of the traditional school system. An extension of the Conservation and Ecology Programme has been a Science Communication Day, held during Science Week. Twenty ecologists, from many Organizations and Universities, communicated to a wide audience of students and teachers from within the education district, and community members. They led hands-on workshops on a diversity of subjects in remnant vegetation, at the beach, at a local wetland and at City Beach Primary School. Topics covered included wetlands, *Banksia/Tuart* Woodlands, coastal rehabilitation and mapping, environmental weeds, creatures of the deep sea, the impact of salt in the Wheatbelt, preparation of an herbarium, what is a leaf?, what is ecology?, the impact of fire, and landform conservation.

A further outcome which has evolved from the development of the bush garden has been the opportunity for ecologists and postgraduate students to communicate to school students and adults. This is a mutually beneficial experience, providing postgraduate students with training in communication skills, valuable for their role as future scientists.

A further outcome has been the development of the school as an important resource centre which hosts inservice courses for teachers and the wider community, provides a venue for Bush Regeneration Courses, supplies advice on bush restoration, community involvement, curriculum materials, educational value of bushland, importance of local species, integration of Conservation and Ecology across the school curriculum, preparation of management plans, nursery establishment, parental involvement, propagation from seed and cutting, recycling to fund conservation and ecology, self-sustaining projects, sponsorship, students club (Bush Guardians) and WaterWise gardens.

Since the development of the Bush Revegetation project the school has been included in a number of publications namely Harris 1994; Keighery and Huston 1994; Powell and Lake 1994; Redreau and Crosby 1996, and City Beach Primary School, Year 6 "The Bush Press", Quarterly Publication.

The quality of the projects and programmes at the school has been recognized by the receipt of a number of awards in the education category, over an extended period of time, as set out below:

1990	John Tonkin Tree Award
1991	State Landcare Award
1991	Represented Western Australia at the National Landcare Awards
1995	John Tonkin Greening Western Australia Award
1995	State Landcare Award
1996	Represented Western Australia at the National Landcare Awards
1996	State Recycling and Waste Reduction Award
1997	Bush Guardians — Finalists in the Young Australian of the Year WA, Unilever Environment Award.

FUTURE DIRECTIONS

The large number of inquiries from other schools, both primary and secondary, regarding the school's Conservation and Ecology Programme indicates the need for a Regional Co-ordinator, situated at the school, to assist others in the establishment of school programmes. As the school has developed a large network of ecologists, postgraduate students, organizations and businesses, the Co-ordinator would be able to link others to this network and support schools wishing to develop programmes. Curriculum Materials developed by the school would also be used to assist those wishing to begin their own school-based projects and programmes.

The development of a Conservation and Ecology Programme at High School level, based on the same principles being employed at City Beach Primary School, would provide an integrated Kindergarten to Year 12 Conservation and Ecology Programme supported by ecologists and facilitated by a Regional Co-ordinator.

CONCLUSION

The success of City Beach Primary School's programmes are the result of implementing scientific advice, and cooperation between staff, parents, students, ecologists, organizations, community members and business. The Case Study has provided methods to incorporate long term action projects and programmes within the constraints of time, finance, personnel, and knowledge which exist within the traditional school system, and to ensure their long-term survival.

The case study has demonstrated the manner in which Science Communication can provide a worthwhile, productive and wide-ranging impetus for introducing sound ecological practices into the wider community.

The outcomes which have arisen from the development of the school's bush garden have supported the preface to the Management Plan (Lake 1991).

> "It is the students' acquired knowledge of identifying and growing local native plants, and their understanding of the environmental issues concerned, that will make a very large contribution."

ACKNOWLEDGEMENT

City Beach Primary School would like to thank all ecologists who have assisted the school including Robert Powell, staff and postgraduate students from the Botany Department, University of Western Australia; Western Australian Museum; Department of Conservation and Land Management; Water Corporation; Ribbons of Blue, Water and Rivers Commission; Kings Park and Botanic Garden; Commonwealth Scientific Industrial Research Organization (CSIRO); Curtin University; Murdoch University; Edith Cowan University; APACE WA; Royal Australasian Ornithologists Union (RAOU); Perth Wildlife Watch; Gould League; and the Local Plants Group.

REFERENCES

City Beach Primary School, Year 6. *The Bush Press*. Quarterly Publication.

Harris, J., 1994. Teaching Among The Trees. Department of Environmental Protection, Perth.

Keighery, B. and Huston, J., 1994. Our Wild Plants. Greening Western Australia, Perth.

Lake, C., 1991. City Beach Primary School Bush Re-Establishment Project Management Plan.

Powell, R. J., 1993. The use of two species of *Parietaria* (Urticaceae) as food plants by the butterfly *Vanessa itea* (Fabricius) in south-western Australia. *Aust. Entomol.* **20**(2): 57–58.

Powell, R., 1996. Growing Locals. Western Australian Naturalists' Club (Inc.).

Powell, R. and Lake, C., 1994. Growing Local Plants For Education. Department Of Conservation and Land Management, Perth.

Redreau, D. and Crosby, D., 1996. Bush Plants For Perth Gardens. National Trust of Australia (WA), Perth.

CHAPTER TWENTY

Conclusion — can we make ecology for everyone?

RICHARD HOBBS[1] and RAY WILLS[2]

THE discussion period at the end of the symposium in Townsville was interesting from a variety of perspectives. The main element of the discussion was an obvious schism between those ecologists who saw a clear need to be involved in communication activities and those who saw such activities as either unnecessary, unwanted or an unwelcome diversion from the real business of being a scientist.

What we are having difficulty with is the fundamental role of scientists in today's world. Several contributors have pointed out that the world is changing and science has to change with it. Picking up on a point from Monica Martinovich, some see it as a question of purity over prostitution. Others view it not as prostitution, but as a necessity, a useful adjunct to what we are doing so that science can actually be useful. In this context Karina Kelly touched on the current backlash against science and scientists. It is a very real phenomenon, perhaps not so much in Australia, but in other parts of the world. Science is becoming a dirty word in many areas and part of the reason for that is that scientists have not been very good at projecting themselves as being relevant and useful to society. The old ivory tower attitude is that we are doing our research and if it is useful then that is good and if not then that is too bad. However, this position is becoming less tenable in today's society. We may not like it but that is the world we are living in.

A question frequently raised is what we do when we are not certain about our facts and in the face of the uncertainty that is inherent in ecological systems. The problem is that we are never certain of our facts, and we are never going to have the "right" answers no matter how much research we do. There is always uncertainty — we live in an uncertain world, we are dealing with an uncertain subject, and we can accommodate that uncertainty into what we say. So we still have to say things because if we wait until we are sure of what we are going to say we will never say anything at all. The problem is an important one, however, especially as more environmental issues end up in the public arena and in the courts. Part of the aim of more effective communication by ecologists needs to be the development of more realistic expectations of how concrete our answers can be. The potential for different interpretations of the same data will not go away, and we need to become cleverer at dealing with this. Economists get away with amazingly inaccurate and fuzzy statements, but ecologists are expected to come up with the "right answer" every time. Getting the balance right between the need for a "bottom line" straightforward answer and the opposing need for hedging with "ifs" and "buts" is an important task for ecologists.

If we accept that ecologists have an increasing requirement to communicate in a variety of spheres, there is still a very valid concern that we are already busting our guts doing what we are supposed to be doing, which is research solving the worlds problems and other great things. If we have to do all this communication stuff on top of it, none of us will have the time, energy and resources to fulfill all our obligations. The answer to that is that we can not all expect to do everything. But as long as everybody does something we would be much further forward than we are now. There are basic techniques to learn which make the task easy, and there are people to help with all this. If we take communication activities on board as being an important and useful part of what we do, we can all contribute in some way.

Let us remember the story of the little girl who went down to the beach after a severe storm, and found millions of starfish that had been stranded on the beach as the tide receded. She started picking up the starfish and taking them back down to the ocean. A man walked along the beach and said to the girl, "There are millions of starfish there — don't you see that your efforts won't really make any difference?" The girl looked at the man, picked up a starfish, ran down to the ocean and dropped it in. She returned with a huge smile on her face and said "Well, it made a difference to that one!"

[1]CSIRO, Division of Wildlife and Ecology, Private Bag, P.O. Wembley, Western Australia, Australia 6014.
[2]Kings Park and Botanic Garden, West Perth, Western Australia, Australia 6005.

P. 110 *in* ECOLOGY FOR EVERYONE: COMMUNICATING ECOLOGY TO SCIENTISTS, THE PUBLIC AND THE POLITICIANS ed by R. T. Wills and R. J. Hobbs. Surrey Beatty & Sons, 1998.

Index